KU-714-308

HUMAN BODY TEMPERATURE

Its Measurement and Regulation

Y. HOUDAS
Lille University
Lille, France

and

E. F. J. RING
Royal National Hospital for Rheumatic Diseases
Bath, England

PLENUM PRESS • NEW YORK AND LONDON

W. SUSSEX INSTITUTE
OF
HIGHER EDUCATION
LIBRARY

Library of Congress Cataloging in Publication Data

Houdas, Yvon.
  Human body temperature.

  Bibliography: p.
  Includes index.
  1. Body temperature—Measurement. 2. Body temperature—Regulation. 3. Fever. I.
Ring, E. F. J. II. Title. [DNLM: 1. Body temperature. WB 270 H918]
QP135.H69 1982   612′.022   82-9856
ISBN 0-306-40872-4

© 1982 Plenum Press, New York
A Division of Plenum Publishing Corporation
233 Spring Street, New York, N.Y. 10013

All rights reserved

No part of this book may be reproduced, stored in a retrieval system, or transmitted,
in any form or by any means, electronic, mechanical, photocopying, microfilming.
recording, or otherwise, without written permission from the Publisher

Printed in the United States of America

Preface

The physiology of man is a complex subject. Unfortunately the regulation of temperature in the human body is not always well explained in textbooks. Many conference proceedings on the subject have been produced that give excellent detail on research topics. However, the subject matter is rarely presented as a composite whole.

New technology has broadened the scope of methods available for studying body temperature. Thermography in particular has made it possible to record in real time the temperature distribution of large areas of the body surface. Modern image processing methods permit dynamic studies to be carried out and detailed analyses made retrospectively—a tremendous advance over the complex and slow techniques formerly used by physiologists. Yet although the association between disease and temperature is as old as medicine itself, beyond the implicit faith in the clinical mercury thermometer, other measuring techniques are finding a slow acceptance.

This book is designed to put into perspective the critical factors that make up "body temperature." Body temperature cannot be viewed as a static entity but rather must be seen as a dynamic process. An understanding of this phenomenon is important to all who use thermal imaging and measuring techniques in clinical medicine. These methods have, in recent years, brought engineers, physicists, technicians, and clinicians together. Inevitably, however, there are gaps and overlaps in technology and understanding. It is our hope that this volume will serve to assist the coordination of the many disciplines with an interest in human body temperature.

The authors wish to acknowledge the assistance of Drs. Ghislaine Carette and Jean-Louis Lecroart (Lille) in the preparation of Chapter 5 and part of Chapter 6, and Dr. R. P. Clark (London) for his constructive advice. AGA Infrared Systems Ltd. (U.K.) and Ultrakust GmbH (F.R.G.) have also supported this work.

Lille, France   Y. Houdas
Bath, U.K.   E. F. J. Ring

Contents

Terminology Used in Thermal Physiology

There is a need for a common language linking physiology to clinical medicine. Certain words and symbols have already been defined.

In physiology the first attempt to propose a glossary of terms was made by a group of workers and published by Gagge *et al.* (1969). Some years later the International Union of Physiological Sciences (IUPS) developed this standard further and the result was a "Glossary of Terms for Thermal Physiology" edited by Bligh and Johnson (1973). This glossary contains a large number of words and symbols, and may therefore be considered almost complete, except on one point: Unfortunately there is a lack of distinction between *heat quantity* and *heat flow* (or *flux*). Respiratory physiologists are used to the distinction between air volume and air flow, and they define air volume by the letter V but air flow (instantaneous flow) by the dotted letter V̇.

In thermal physiology a similar distinction must be made. *Heat quantities* are measured in joules. They may also be related to time and then measured in joules per minute. This expression is $\Delta H / \Delta t$. *Heat flux* is mathematically the time derivative of a quantity of air. This is expressed as dH/dt and expressed in watts. For this reason, dotted letters are used in the following discussion to express heat flux.

On the other hand, the IUPS glossary proposes three different symbols for expressing the three different pathways of the same basic phenomenon, i.e., heat flux: convection, conduction, radiation. Physicists use only one symbol, usually the Greek letter Φ. It would be more appropriate to use the same letter (or the letter $\dot{H}$) with a subscript indicating the pathway of transfer (as used by Cena and Clark, 1981).

Finally, and probably because the IUPS glossary was made by pure physiologists, it does not define the terms used in medical thermology. For this reason, the European Association of Thermology (EAT) has created a Commission for defining those terms used especially in medical practice. (The authors are involved in these organizations.) This work is also concerned with the terms defining specific devices and techniques and has been published in "Thermographic Terminology" (1978).

We have followed this terminology for medical applications and the IUPS terminology for the physiological point of view, with the addition (mentioned above) of the symbols for heat flux.

1.1. UNITS

1.1.1. Heat Quantity

☐ The *calorie* (cal) is the quantity of heat needed to increase the temperature of one gram of water from 15°C to 16°C. The third power—the *kilocalorie* (kcal)—is generally used. 1 kcal = 1000 cal.

☐ However, the unit recommended by the Système International d'Unités (SI) is the *joule* (J), which is the quantity of energy corresponding to the work of a force of one newton moving one meter. The unit commonly used is the *kilojoule* (kJ).

☐ Equivalence: 1 cal = 4.18 J
            1 kcal = 4.18 kJ

1.1.2. Heat Flux

☐ The kilocalorie per hour (kcal/h) must be avoided.
☐ The official unit is the *watt* (W), which is the power corresponding to 1 joule per second.

- ☐ Equivalence: $1 \text{ kcal} \cdot \text{h}^{-1} = 1.16 \text{ W}$
- ☐ Note that the symbol generally used by physicists is the Greek letter Φ. However, as described above, the concept of heat flux has not been correctly considered by the IUPS panel.

1.1.3. Temperature

- ☐ The *degree Celsius* (°C) is the hundredth part of the difference between the temperature of melting ice (0°C) and that of boiling water (100°C) at the standard pressure of 1013 millibars. (Note that the term *degree centigrade* should be avoided.)
- ☐ The *Kelvin* (K) has the same value as the degree Celsius but zero corresponds to -273.5°C. (Note that the unit is the Kelvin and not the *degree* Kelvin, and that the symbol is K and not °K.)

1.1.4. Pressure

- ☐ The SI unit is the *pascal* (Pa), which is the pressure developed by 1 newton on 1 square meter: $1 \text{ Pa} = 1 \text{ N} \cdot \text{m}^{-2}$.
- ☐ Meteorologists generally use the *millibar* (mb) and physiologists the *millimeter of mercury* (mm Hg) or the *torr* (torr). The *atmosphere* (atm) is also used.
- ☐ Equivalence:
$$1 \text{ bar} = 10^5 \text{ Pa}$$
$$1 \text{ mb} = 10^2 \text{ Pa}$$
$$1 \text{ atm} = 1013 \text{ mb} = 760 \text{ mm Hg}$$
$$1 \text{ torr} = 1/760 \text{ atm}$$
$$1 \text{ mm Hg} = 1.00000014 \text{ torr}$$
and especially
$$1 \text{ mm Hg} = 133.3 \text{ Pa}$$

1.2. PHYSICAL QUANTITIES

1.2.1. Basic Quantities

a   Thermal diffusion, expressed by the ratio $k/\rho c$. Units: $\text{m}^2 \cdot {}^\circ\text{C}^{-1} \cdot \text{s}^{-1}$.

α Absorbance, a nondimensional number that can vary from 0 to 1.

β Dilation coefficient (thermal expansion coefficient) of a fluid. Unit: K^{-1}.

c   Specific heat. Units: $J \cdot g^{-1}$. (As a subscript, denotes "convective.")

c_p For a fluid, specific heat at constant pressure.

E Internal energy of a system. Unit: J.

ε Radiant emissivity, a nondimensional number that can vary from 0 to 1.

G Free energy of a system. Unit: J.

g   Value of terrestrial gravitational acceleration. Units: $m \cdot s^{-2}$. Value: $9.81 \ m \cdot s^{-2}$.

H Enthalpy of a system. Unit: J.

k   Thermal conductivity. Units: $W \cdot m^{-1} \cdot {}^\circ C^{-1}$. (As a subscript, denotes "conductive.")

λ Wavelength of electromagnetic radiation. Unit: 10^{-6} m or μm.

λ Latent heat of vaporization of water. Units: $J \cdot g^{-1}$.

μ Viscosity of a fluid. Unit: poise (P).

ν Dynamic viscosity of a fluid expressed by the ratio ρ/μ. Units: $g \cdot cm^{-2} \cdot Pa^{-1}$.

ρ Density or specific mass. Units: $g \cdot cm^{-3}$.

S Entropy of a system. Units: $J \cdot K^{-1}$.

σ Stephan–Boltzmann constant, $56.7 \times 10^{-9} \ W \cdot m^{-2} \cdot K^{-4}$.

U Speed of displacement of a fluid. Units: $m \cdot s^{-1}$.

kρc   "Thermal inertia." Units: $J^2 \cdot cm^{-4} \cdot {}^\circ C^{-1} \cdot s^{-1}$.

1.2.2. Quantities Used in Physiology

Apart from the particular symbols listed below, the IUPS recommends the use of physical symbols adding a subscript to indicate the site or conditions of measurement. For instance, the symbol for temperature is T. Thus ambient temperature can be symbolized T_a and rectal temperature, T_{re}.

A   Surface area of heat transfer. Unit: square meter (m^2). Used generally with a subscript indicating the route of transfer, for instance, A_R for surface of radiative heat transfer.

C Heat transferred by convection. Unit: W (generally per m^2).

E Heat lost by evaporation. Unit: J.

$\dot{E}$ Heat flux by evaporation (evaporative heat rate). Unit: W (generally per m^2).

H In physics, the symbol for enthalpy. In physiology, often used to symbolize heat flow rate. Unit: W (generally per m^2). Subscripts are frequently used to distinguish different meanings.

$\dot{H}_e$ Sum of the heat flow rates transferred by conduction, convection, and radiation (sensible heat transfer).

$\dot{H}_i$ or $\dot{H}_m$ Rate of metabolic heat flow.

h   Heat transfer coefficient. A subscript indicates the route of transfer, for instance, h_k for conductive heat transfer coefficient and h_c for convective heat transfer coefficient. Units: $W \cdot m^{-2} \cdot °C^{-1}$.

M Metabolic free energy production, which corresponds to the total energy produced by the body and includes all forms of energy. As it is expressed as a rate of energy production, the appropriate symbol is $\dot{M}$, expressed in $W \cdot m^{-2}$ of surface area. $\dot{M}$ is generally determined by measuring oxygen consumption.

ṁ   Sweat secretion rate. Units: g per time unit.

P   Partial pressure of a gas, for example, Po_2 for partial pressure of oxygen. For water vapor pressure, the symbol is generally P_w for simplicity. Unit: mm Hg or torr.

S Heat storage. Unit: J.

$\dot{S}$ Rate of heat storage, the difference between the rate of heat gain and the rate of heat loss. Unit: W (generally per m^2).

W More accurately $\dot{W}$, that part of energy appearing as mechanical energy during muscular exercise. Unit: W (generally per m^2).

1.3. SOME COMMON DEFINITIONS

Area, total body (A_b) The area of the outer surface of a body, assumed to be smooth, in m^2 [IUPS]. Generally determined by the Dubois formula, relating it to the weight (W, in kg) and height (H, in m) of the body:

$$A_b = 0.202 \ W^{0.425} H^{0.725}$$

Basal metabolic rate   The rate of metabolic free energy production, calculated from measurements of the heat production or
oxygen consumption of an organism in a rested, awake, fasting,
and thermoneutral state [IUPS].

Blackbody   An ideal body that absorbs all electromagnetic radiation falling on its surface or, when it is acting as a radiator, whose
surface radiation can be described by Planck's law [EAT].

Body heat balance   The steady state relation in which rate of heat
production in the body equals rate of heat loss to the environment
[IUPS].

Body heat content   The product of the body mass, the average
specific heat, and the absolute mean temperature (in J) [IUPS].

Calorimetry   In physiology, the measurement of heat transfer
between a tissue, an organ, or an organism and its environment.

Conduction, heat transfer by   Transfer of heat within a solid or
between two solids.

Convection, heat transfer by   Transfer of heat between a solid and
a fluid, or between two fluids, or within a fluid.

Core temperature or deep body temperature   The mean temperature
of the tissues at a depth below that which is affected directly by a
change in the temperature gradient through peripheral tissues
[IUPS].

Effective temperature (T_{eff})   An arbitrary index that combines in a
single value the effects of temperature, humidity, and air movement on the sensation of warmth or cold felt by human subjects.
The numerical value is that of the temperature of "still" air (in °C)
saturated with water vapor that would induce an identical sensation.

Humidity, air   *Absolute*: The mass of water vapor in a unit volume
of air (in kg · m^{-3}). *Relative*: The ratio of the mass of water vapor
actually present in a unit volume to the mass of vapor required to
saturate that volume at the same temperature.

Hyperthermia   The condition of a temperature-regulating animal
in which the core temperature is more than one standard deviation

above the mean core temperature of the species in resting conditions in a thermoneutral environment.

Hypothermia   The condition of a temperature-regulating animal in which the core temperature is more than one standard deviation below the mean core temperature of the species in resting conditions in a thermoneutral environment.

Neutrality, thermal, or neutral zone, thermal   The range of ambient temperature within which metabolic rate is at a minimum, and within which temperature regulation is achieved by nonevaporative physical processes [IUPS].

Pyrogen   The generic term for any substance, whether exogenous or endogenous, that causes a fever when introduced into or released within the body [IUPS].

Radiant heat exchange   The net rate of heat exchange by radiation between an organism and its environment.

Shivering, thermogenesis by   An increase in the rate of heat production during cold exposure owing to increased contractile activity of skeletal muscles not involving voluntary movements and external work.

Storage of body heat, rate of   The rate of increase ($+$) or decrease ($-$) in the heat content of the body caused by an imbalance between heat production and heat loss ($\dot{S}$ in W).

Sweating, thermal   A response of the sweat glands to a thermal stimulus.

Temperature reference source   Any medium of known temperature suitable for calibration. In infrared thermography, a temperature reference may be any radiator of known temperature whose emissivity is known and constant within a given wavelength range, and is suitable for temperature calibration (this term does not necessarily imply that the radiator or source is a black body) [EAT].

Thermal comfort   Subjective satisfaction with the thermal environment [IUPS].

Thermography   The recording of the temperature or temperature distribution of a body (whether obtained by conduction, convection, or radiation) [EAT].

Thermography, contact   The recording of the temperature or temperature distribution of a body when the thermal sensor is in contact with that body [EAT].

Thermography, infrared   The recording of the temperature or temperature distribution of a body using infrared radiation emitted by the surface of that body at wavelengths between 0.8 μm and 1.0 mm [EAT].

Thermography, microwave   The recording of the temperature or temperature distribution of a body using the microwave energy emitted by that body at wavelengths between 1 mm and 1 m [EAT].

Principles of Heat Transfer

2.1. THE SYSTEM AND ITS ENVIRONMENT

A *system* is any part of the universe that can be studied. It may be an atom or a living body, the earth or the solar system. All its surrounding matter constitutes the *environment* of the system.

If the system exchanges matter with its environment, it is called an *open system*. If there is no matter exchange, it is called a *closed system*. If neither matter nor heat are exchanged, it is said to be a *thermally insulated system*.

2.2. ENERGY

Energy can be defined as the ability to do work. It exists in many forms, such as mechanical, electrical, nuclear, and thermal.

Transformation of one form of energy into another is possible, for example, nuclear energy can produce electricity. Such transformation is never total, and the part of nuclear energy that cannot be transformed into electricity appears as thermal energy (heat). The transformation of all these forms of energy into heat can be total, but the contrary, i.e., the transformation of heat into one of the other forms, is never complete. Heat thus appears to be a particular form of energy because it is the ultimate form of energy resulting from transformation. It is therefore possible to consider it as "physical waste."

Units of energy differ according to their form. For instance, the unit of thermal energy is the *calorie* (cal), the amount of heat needed to increase the temperature of 1 g of water from 14.5 to 15.5°C. On the other hand, the unit of mechanical energy is the *erg*, and that of electrical energy, the *joule* (J). However, as will be seen in the following section, there is an equivalence between all forms of energy.

2.3. FIRST LAW OF THERMODYNAMICS

2.3.1. Equivalence between Different Forms of Energy

The *first law of thermodynamics* states that the various forms of energy can be changed from one form to another, but cannot be created or destroyed. If a system has received, say, mechanical energy, W, and if its internal energy, U, has not varied, it has necessarily lost an equivalent amount of some other energy, e.g., heat, Q. There is a balance between the energy gained, which is counted positively, and the energy lost, which is counted negatively:

$$W = -Q$$

If the internal energy of the system has varied from the value U_1 to the value U_2, it is possible to write the following generalized equation:

$$U_2 - U_1 = W + Q$$

Although Lavoisier, as early as 1783, made the assumption of the equivalence of energy, the first law of thermodynamics was actually first expressed in 1842 by Joule, who showed experimentally the equivalence between mechanical work and thermal energy.

Table 2.1 shows the main units of energy and their equivalents. Because of these equivalents, it is now internationally accepted that the joule (J) must be used as the unit of energy, whatever the form.

Table 2.1. Units of Energy and Their Equivalents

	Form of energy		
	Mechanical	Thermal	Electrical
Dimension:	Force × distance (dyne × cm)	Temperature × specific heat	Potential × current (volt × ampere)
Unit and symbol:	erg	calorie (cal)	joule (J)
Equivalents:	1 erg =	$2.389 \cdot 10^{-8}$ cal =	1.10^{-7} J
	$418 \cdot 10^{5}$ ergs =	1 cal =	4.18 J
	10^{7} ergs =	0.24 cal =	1 J

This unit is small, and its multiple, the kilojoule (kJ), is more often used, especially in studies of thermoregulation in man.

2.3.2. Release of Potential Energy in Living Tissues

Animals find energy in foods in which it has been stored. In the cell, this energy is released by oxidation–reduction reactions. The term *oxidation* conveys the impression of a reaction involving combination with oxygen:

$$(CH_2O)_n + nO_2 \rightarrow nCO_2 + nH_2O + \text{Energy}$$

In this example, the molecules of carbohydrates are oxidized by the addition of oxygen atoms. However, oxidation is more generally considered as the removal of hydrogen atoms or electrons from a molecule.

Conversely, *reduction* is the addition of hydrogen atoms to a molecule, as in the formation of ammonia:

$$N_2 + 3H_2 \rightarrow 2NH_3$$

In biological reactions, a reduction occurs for every oxidation reaction because the hydrogen atoms released by oxidation do not remain free. They are immediately tied up by other atoms or molecules. Thus the general concept of energy release can be represented as follows:

AH_2	+	B	→	A	+ BH_2 +	Energy
Substance to be oxidized (potential energy)		Substance to be reduced (generally O_2)		Substance oxidized (generally CO_2)	Substance reduced	To be captured and/or transformed

2.3.3. Enthalpy

Consider a system in which a chemical reaction occurs. The initial internal energy is U_1 and the final internal energy is U_2. If the reaction is carried out at constant volume (no gas expansion), there is no mechanical work. Therefore

$$U_2 - U_1 = Q_v$$

where the subscript v indicates that the reaction occurs at constant volume. If Q_v is positive, the reaction is said to be *endothermal*. If Q_v is negative, the reaction is *exothermal*. The combustion of food in a calorimetric bomb is the classic example of an exothermal reaction at constant volume in a thermally insulated system (*adiabatic reaction*).

In living bodies, reactions generally occur at constant pressure. It is thus possible to write

$$U_2 - U_1 = W + Q_p$$

For a gas, the mechanical energy W corresponds to the product of pressure and volume, $P \cdot V$. Thus

$$W = -P(V_2 - V_1)$$

which then becomes

$$U_2 - U_1 = PV_1 - PV_2 + Q_p$$

or

$$Q_p = (U_2 + PV_2) - (U_1 + PV_1)$$

The term $U + PV$ is called the *enthalpy* and is symbolized by H. Enthalpy has the dimension of energy. Therefore

$$Q_p = H + W$$

For example, the complete oxidation of 1 mole of D-glucose at 25°C and 760 mm Hg produces 673 kcal (2813 kJ) according to the following reaction:

$$C_6H_{12}O_6 + 6O_2 \rightarrow 6CO_2 + 6H_2O - 673 \text{ kcal}$$

Note that the negative sign is used before 673 kcal because heat is given off by the system (exothermal reaction).

In this reaction, there is no change in gas volume. Six liters of O_2 are used and 6 liters of CO_2 are produced. Therefore, $W = 0$ and

$$\Delta U = Q_v = \Delta H$$

For these kinds of chemical compounds, ΔH can be measured by the combustion of the substance in a calorimetric bomb. Since the energy given off by this combustion is the same whatever the route of oxidation, ΔH is the same in the calorimetric bomb as in the living body. Tables exist giving the values of ΔH for the most common compounds that can be oxidized in the body (see Tables 2.2 and 2.3).

Table 2.2. Energy (in kJ) Produced by the Oxidation of 1 g of Three Substances[a]

Substance	Total energy of oxidation	Energy furnished by metabolism in living bodies	Effective value of energy, after taking account of intestinal absorption
Glucose	17.55	17.55	17.15
Lipid	39.30	39.30	37.60
Protein	23.40	19.85	17.15

[a] Note that this energy cannot be totally used by the living body, because intestinal absorption does not affect all molecules equally. Moreover, proteins are not completely oxidized (urea is the most common waste of this group of compounds).

Table 2.3. Energy (in kJ/100 g) Produced by the Oxidation *in Vivo* of Some Alimentary Products[a]

Product	Energy produced
Animal product	
Meat (except fatty tissues, such as bacon)	975
Poultry, fish	555
Eggs	610
Milk	290
Butter	3030
Cheese	1255
Vegetable products	
Leafy vegetables	55
Green and yellow vegetables	140
Tomatoes	110
Dry beans and peas	1785
Potatoes	305
Citrus fruits	160
Other fresh fruits	200
Flour and cereal products	1530
Bread	1150

[a] Mean values.

2.4. SECOND LAW OF THERMODYNAMICS

When a warm object is put into cold water, its temperature falls, but the water temperature increases, until the two become equal. This is due to energy transfer from the high level (high temperature) to the low level (low temperature). However, the reverse does not occur spontaneously. An analogy can be drawn with the mechanical energy produced in a waterfall. Water goes down from one level to a lower one because gravitational energy is higher at the upper level. In this case the reverse transfer of energy cannot occur spontaneously.

The first law of thermodynamics expresses the equivalence between the various forms of energy, but it cannot predict the spontaneous evolution of a reaction or an energy exchange. This prediction is the purpose of the *second law of thermodynamics*, first proposed in 1824 by Carnot. In its present form, this law states that energy transformation can occur spontaneously only in one sense, that is, toward greater entropy.

Entropy is a physical concept that is not easy to describe in simple language. If we assume a reversible transformation, in which the system exchanges thermal energy Q with its environment at temperature T, the entropy of the system varies from S_1 to S_2 and this variation can be written

$$S_2 - S_1 = \Delta S = Q_{rev}/T$$

(The subscript rev means that the reaction is reversible.) This case is ideal. In practice, the reactions are not reversible; in this case ΔS is greater than Q_{irr} (the subscript irr means that the reaction is irreversible).

If we now consider an insulated system in which a reversible reaction has occurred, entropy has not varied because $Q_{rev} = 0$. Thus we have $\Delta S = 0$. If the transformation occurs spontaneously, we now have $Q_{irr} = 0$. This necessarily implies that ΔS is positive.

This conclusion is very important: *The only possibility for spontaneous variation of the entropy of a closed insulated system is its increase.*

If we consider again the example of a warm object put into cold water, we see that the maximum entropy will correspond to thermal equilibrium. The same reasoning can be applied to all other energy transformations. Total equilibrium always corresponds to the maximal entropy. Therefore, the gradual evolution of a living body into physical, chemical, and thermal equilibrium with its environment is death.

On the contrary, life, which is the creation of new molecules with the storage of potential energy, corresponds to negative entropy. This is physically possible only because the living body is not a closed insulated system. It receives energy from the sun, either directly or indirectly.

Free Enthalpy, Usable Energy. When a molecule of food is oxidized, a certain amount of energy depends only on the initial state (the molecule of food) and its final state (the waste molecules), but is independent of the intermediate states. We have seen that the amount of energy produced could be considered as a variation in the enthalpy H. However, this energy cannot be completely transformed into mechanical, chemical, electrical, or other types of work. The major part of this energy must appear in the form of heat and corresponds to the variation of entropy.

Free enthalpy or *free energy* is that fraction of energy that can be transformed into work. It is symbolized by G (or ΔG) and is defined according to the Gibbs equation as follows:

$$\Delta G = \Delta H - T\Delta S$$

As ΔS is always positive (except in an ideal case), $T\Delta S$ is also always positive, except at absolute zero. Therefore, all living bodies produce heat, which can be considered as "physical waste."

On the other hand, if $T\Delta S$ is positive, this implies that ΔG is greater than ΔH (in absolute values). Such reaction is termed *ex-oergic* because is can supply the energy for work. For example, in the complete oxidation of 1 mole of D-glucose at a temperature of 298 K, ΔG is -686.48 kcal and ΔH is -673.00 kcal. Entropy has then been increased by $(686.48 - 673.00)/298$, i.e., 45.2 e.u.* On the contrary, the synthesis of 1 mole of D-glucose corresponds to a negative entropy ($T\Delta S = -13.480$ kcal). This reaction is said to be *endoergic* and can occur in living bodies only because an equivalent amount of energy has been supplied by the environment.

In the thermal function of homeotherms, only that part of energy which appears as heat is directly involved. For this reason, the next section will show how heat is exchanged within the body and/or between the body and its environment.

2.5. THE TRANSFER OF HEAT

The study of heat transfer is called *thermokinetics.*

When there is a difference of temperature between two bodies or between parts of the same body, heat flows from hot areas to cold areas spontaneously so that the temperature difference between them is diminished toward zero. Therefore, if a warm body at temperature T_b is put into a cooler environment at temperature T_a, the body will progressively cool. If the mass of the body is small with respect to

---

*e.u. = Entropy unit, expressed in kcal/K.

that of its environment, T_b will decrease until it reaches the value of T_a. The variation of T_b is described by Newton's law, first published in 1701:

$$dT_b/dt = k_n(T_b - T_a)$$

where k_n is Newton's cooling constant.

A temperature variation thus occurs because heat has been transferred from the body to its environment. This heat transfer depends directly on the temperature difference between the body and its environment. When T_b becomes equal to T_a, there is no further heat transfer. The system is then in thermal equilibrium. If the body itself is producing heat continuously, T_b always remains higher than T_a. If this temperature difference does not vary according to time, the system is said to be in a *steady state*.

Two possibilities for heat transfer are described in physics:

1. *Heat transfer without change of state.* This can occur by either conduction, convection, or radiation.
2. *Heat transfer with change of state.* For example, to achieve conversion of water from liquid to vapor by boiling requires heat transfer, although the water may remain at a constant temperature.

2.5.1. Heat Transfer without Change of State

2.5.1.1. Conduction

Heat transfer by *conduction* occurs within a solid or between two or more solids in close contact.

For example, if a metal bar has one end at temperature T_2 and the other at temperature T_1, in a steady state heat flows from the warm end (T_2) to the cold end (T_1). The heat flow rate $\dot{Q}_k$ throughout a transverse section of area A_k is described by the relationship proposed by Fourier in 1822:

$$\dot{Q}_k = -(k/x)(T_2 - T_1)A_k$$

where k is the thermal conductivity of the material and x is the distance between the ends of the bar.

In thermal physiology, the ratio k/x is called the *conductive heat transfer coefficient*, expressed in $W \cdot m^{-1} \cdot °C^{-1}$ and symbolized by h_k. The ratio k/x is expressed in $W \cdot m^{-2} \cdot °C^{-1}$. Its inverse ratio x/k is then given in $m^2 \cdot °C \cdot W^{-1}$, and the ratio $x/k \cdot A_k$ in $°C \cdot W^{-1}$ represents *thermal resistance* and can be symbolized by R_k. It is thus possible to express heat transfer by conduction using a formula similar to Ohm's law, which is well known in electricity:

$$\dot{Q}_k = (1/R_k)T \qquad\qquad I = (1/R)V$$

Heat transfer   Ohm's law

The concept of thermal resistance (or *impedance*) corresponds to that of *thermal insulation* and is very useful in practice because it can be measured. When heat flows through different structures or materials with different thermal insulations (for example, a clothed subject), the total thermal resistance and thus the total conduction coefficient can be calculated.

Applying the analogy with Ohm's law, the total thermal resistance of n layers having respective resistances of $R_{k_1}, R_{k_2}, \ldots, R_{k_n}$ is

$$R_k = R_{k_1} + R_{k_2} + \cdots + R_{k_n}$$

On the other hand, in physiology, the problem of heat transfer by conduction is more complex, because (1) a steady state is rarely achieved, (2) heat is produced directly by the body, and (3) transfer occurs in three spatial dimensions.

Consider a small cube within a body. At any given point in this cube, the temperature can be related to the three coordinates, x, y, and z. The heat flowing into one face, for instance that corresponding to the x axis, is described by the following equation:

$$\dot{Q}_x = -k\frac{dT}{dt}(dy, dz)$$

The heat flowing out from the opposite surface is $\dot{Q}_x + dx$, the difference corresponding to the heat produced by the cube and to the external variation in temperature of the cube.

If a similar reasoning is applied to all three axes, the final equation is

$$\frac{d^2T}{dx^2} + \frac{d^2T}{dy^2} + \frac{d^2T}{dz^2} + \frac{\dot{Q}_0}{k} - \frac{\rho c}{k} \cdot \frac{dT}{dt} = 0$$

Heat flowing throughout the three faces	Heat produced	Variation in temperature

The inverse of the ratio $\rho c/k$ is generally termed *thermal diffusion* and symbolized as a.

This equation expresses the fact that the sum of the heat entering the three faces during time dt, added to the spontaneous heat production during the same time, is equal to the sum of heat dissipation from the three faces, plus the algebraic variation in heat content. This is of fundamental importance because all other equations can be derived from it by mathematical procedures, e.g., differentiation and integration.

For instance, if no heat is produced, the term $\dot{Q}_0$ equals zero and the equation becomes

$$\frac{d^2T}{dx^2} + \frac{d^2T}{dy^2} + \frac{d^2T}{dz^2} = \frac{1}{a} \cdot \frac{dT}{dt}$$

This is the general expression for Fourier's law.

Another example: If there is no heat source and if the system is in steady state, the repartition of temperatures within the cube follows Laplace's equation:

$$\frac{d^2T}{dx^2} + \frac{d^2T}{dy^2} + \frac{d^2T}{dz^2} = 0$$

In practice, the equations given in this chapter are used especially to determine the thermal characteristics and thermal production of tissues, particularly when pathological, such as a tumor.

2.5.1.2. Convection

Heat transfer by *convection* occurs in a fluid volume or between a fluid and a solid, providing a temperature gradient exists. Convection differs from conduction in that the fluid itself can be moving, and thus heat can be transferred not only by "contact" between the molecules, but also by their displacement. If molecular movement is due only to density differences produced, for example, by temperature differences within the fluid, convection is said to be *natural*. In all other conditions, it is described as *forced*. The greater the speed of the fluid in relation to a solid, the greater will be the rate of heat transfer by convection.

The general law for this form of heat transfer is similar to that for conduction and can be written

$$\dot{Q}_c = -h_c \Delta T A_c$$

where $\dot{Q}_c$ is the heat flux exchanged by convection, h_c is the *convective heat transfer coefficient* or *convection coefficient*, ΔT is the temperature difference between the fluid and the solid, and A_c is the surface area transferring the heat.

Convection differs from conduction mainly in the nature of their respective coefficients h_c and h_k. Whereas h_k is dependent principally on the thermal characteristics of the fluid and solid, h_c also depends on such factors as the fluid density ρ and the relative speed of displacement U.

2.5.1.2.1. Natural Convection. In *natural convection* fluid molecules are moved or displaced only by the influence of density differences created by temperature changes. Therefore, the convection coefficient h_c is related to the factors involved in this movement:

The heat conductivity of the fluid, k
Its viscosity, μ
Its specific mass (density), ρ
Its dilation coefficient (thermal expansion coefficient), β
Its specific heat (at constant pressure), c_p

These factors can be grouped into three adimensional numbers*:

The Nusselt number, $\mathrm{Nu} = h_c L/k$
The Prandtl number, $\mathrm{Pr} = \mu c_p/k$
The Grashof number, $\mathrm{Gr} = g\beta\nu^2$

In these numbers, L is a geometric characteristic of a body, such as its length (l) or its diameter (d), g is the gravitational acceleration ($9.81 \ \mathrm{m} \cdot \mathrm{s}^{-2}$), and ν is the *kinematic viscosity*, which is the ratio μ/ρ.

However, whatever the geometric shape of the body and the nature of the fluid, the experimental results show that Nu is a function of Pr and Gr:

$$\mathrm{Nu} = a(\mathrm{Pr} \cdot \mathrm{Gr})^m$$

where a and m are constants that can be determined experimentally. Thus, if the above factors and their relationships are known, h_c can be calculated, e.g., in high- or low-pressure atmospheres, or in gas mixtures that differ from air.

2.5.1.2.2. Forced Convection. In *forced convection*, h_c is mainly dependent on air displacement. Therefore, a new adimensional number is needed to take account of this factor, as the Grashof number relating to free movement of molecules is inapplicable. This new number includes the rate of relative air displacement, U, and is expressed by the ratio

$$\mathrm{Re} = UL/\nu$$

where Re is the *Reynolds number* and L and ν have previously been defined.

However, as h_c is dependent on fluid movement, it is also influenced by the patterns of movement and varies according to whether the fluid displacement is laminar or turbulent. In *laminar flow*, Re is low, but in turbulent flow it is high, an approximate transitional zone corresponding to Re values between 2000 and 10,000. Figure 2.1 illustrates the patterns of laminar and turbulent flow.

Adimensional numbers are products or ratios that are expressed without units, in which all constitutive parameters are given in a homogeneous unit system.

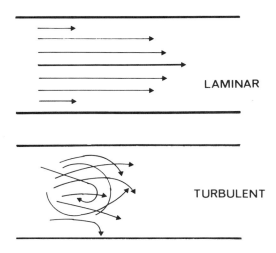

Figure 2.1. Patterns of fluid flow.

a. Laminar Flow. In laminar flow the fluid can be represented by equal layers sliding easily one on the other. However, if one considers laminar fluid flow along a solid plane, the speed of fluid displacement is not the same close to the surface of the plane as it is at a distance from the surface. The distribution of speed between the value 0 at the interface and the value U at a certain distance x is represented in Figure 2.2. This flow can be divided into two parts—an

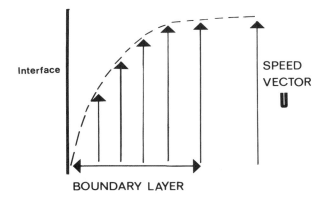

Figure 2.2. The formation of a boundary layer in a laminar flow system.

external part, where fluid moves homogeneously at speed *U*, and another part, consisting of a thin layer along the surface within which there is a speed gradient. This layer was called the *boundary layer* by Prandtl at the beginning of the century.

b. Turbulent Flow. Turbulent flow is characterized by the existence of local alterations in the flow. Disturbances occur from one point to another, and at a given point from one time to another; therefore, at a fixed point there are speed changes both in intensity and direction. Nevertheless, it can also be assumed that the fluid is composed of at least three successive layers (Figure 2.3):

1. The flow itself, which is turbulent.
2. A basal laminar flow in close contact with the surface.
3. A barrier layer between the two flows.

In this case, the basal laminar layer is very thin.

c. Concept of Thermal Boundary Layer. By analogy with the concepts of laminar or turbulent flow and of the dynamic boundary layer, it is possible to describe a *thermal boundary layer* in which there is an important temperature gradient between the solid surface and the fluid at a certain distance. At the interface, heat transfer occurs by conduction between the molecules of the solid and those of the fluid. In the fluid itself, heat transfer occurs by true convection. However, the term *convection* and the related convection coefficient cover the whole phenomenon.

The patterns of flow modify the thickness of the thermal boundary layer and consequently the value of h_c. However, dynamic

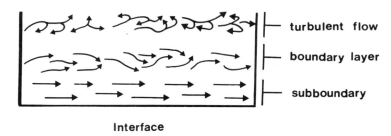

Figure 2.3. Patterns of flow in a turbulent flow system.

boundary layers and thermal boundary layers are not of the same thickness. The dynamic interface is thicker than the thermal interface if Prandtl's number is greater than 1, and the inverse is also true. In air, Pr equals 0.72, and the thermal boundary layer is thus thicker than the dynamic boundary layer.

2.5.1.2.3. Values of h_c in Forced Convection. As in natural convection, h_c can be calculated if Nu is known. Here Nu depends on the characteristics of the fluid expressed by Prandtl's number and also on the patterns of air motion described by the Reynolds number. The general equation is

$$Nu = a_1(Re^n Pr^m)$$

where a_1, n, and m are constants. When the fluid is air, Pr is given and the equation may be simplified as follows:

$$Nu = a_2 Re^p$$

The constants a_2 and p are different according to whether the flow is laminar or turbulent. They can be determined experimentally. For instance, for a cylinder subjected to air flowing perpendicularly to its axis, one has

$$Nu = 0.5 Re^{0.47}$$

if Re is between 10 and 10^3 (laminar flow), and

$$Nu = 0.18 Re^{0.63}$$

if Re is between 10^4 and 10^5 (turbulent flow). The first value is close to that found by Rapp (1971) for a prone subject placed in air flowing slowly and perpendicular to its axis:

$$Nu = 0.595 Re^{0.5}$$

2.5.1.3. Radiation

All bodies emit electromagnetic radiation (Figure 2.4) to a greater or lesser degree. Certain electromagnetic waves are capable of

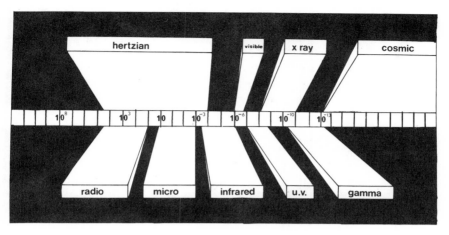

Figure 2.4. The electromagnetic spectrum.

transporting large quantities of thermal energy. These are essentially long-wave radiation with wavelengths between 10^{-7} and 10^{-4} m. This includes all visible radiation of wavelengths 4.10^{-7}–8.10^{-7} m, as well as most of the infrared and part of the microwave range. The curve representing the emission of a homogeneously radiating body versus the whole wavelength range is called the *spectroradiometric curve* (Figure 2.5).

2.5.1.3.1. Concept of a Blackbody. A *blackbody* is defined as an object that absorbs all radiations falling upon it, at any wavelength. Kirschoff's law states also that such a body is equally capable of emitting all radiation.

Blackbody emissivity. The emissivity of a blackbody can be described by three equations:

1. *Planck's law.* This general law describes the emitting power, $W(\lambda)_b$, of a blackbody within a spectral interval of 1 μm at wavelength λ:

$$W(\lambda)_b = \frac{2\pi hc^2}{\lambda^5 \exp(hc/\lambda kT) - 1}$$

where h is Planck's constant (6.6×10^{-34} W · s^{-2}), k is Boltzmann's

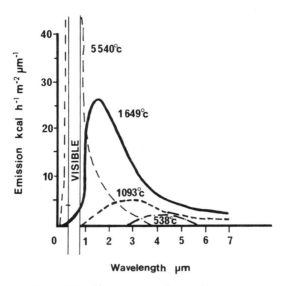

Figure 2.5. The spectroradiometric curve.

constant $(1.4 \times 10^{-23}$ J $\cdot$ K^{-1}), T is the absolute temperature (in K) of the blackbody, and c is the velocity of light $(3 \times 10^{10}$ cm/s^{-1}).

2. *Stefan–Boltzmann's law.* This law expresses the total energy emitted by a blackbody, W_b. It was formulated experimentally by Stefan in 1879 and confirmed theoretically by Boltzmann in 1884. It corresponds to the integration of Planck's law from $\lambda = 0$ to $\lambda = \infty$:

$$W_b = \sigma T^4$$

where σ is the Stefan–Boltzmann constant $(56.7 \times 10^{-9}$ W $\cdot$ m^{-2} $\cdot$ K^{-4}). This law is very important in practical infrared thermography because it states that the total emissive power of a blackbody is proportional to the fourth power of its absolute temperature.

3. *Wien's displacement law.* This law expresses the fact that the wavelength at which the emissive power of a blackbody is maximum varies inversely with the absolute temperature of the body. It corresponds to the derivative of Planck's law with respect to λ:

$$\lambda_{max} = 2898/T \qquad [\text{in } \mu\text{m}]$$

This law is the mathematical expression of a common observation: As the temperature of, for instance, an iron bar increases, its color varies progressively from blue-black to red and yellow, owing to a decrease of the main wavelength of the radiative emission. At room temperature, T is about 300 K and the peak of radiant emittance of common objects lies at 9.7 μm, that is, the far infrared.

2.5.1.3.2. Nonblackbody Emitter (Graybody). Real objects rarely comply with these laws over an extended wavelength region, although they may approach the characteristics of a blackbody in certain spectral intervals. More typically, when incident radiation falls on a real object, it can be partly absorbed (α), reflected (ρ), or transmitted (τ) (Figure 2.6). As we have seen, all these factors are more or less wavelength-dependent. Therefore, the symbol λ will be used to imply the spectral dependence of α, ρ, and τ.

The sum of these three fractions at any wavelength must be equal to the incident radiation. If the value 1 is attributed to the incident radiation, we have the equation

$$\alpha(\lambda) + \rho(\lambda) + \tau(\lambda) = 1$$

On the other hand, graybodies are capable of radiating energy. However, at a given temperature, a graybody always emits less than a blackbody, and its radiant emissivity can be defined by the fraction of its blackbody emissivity. Mathematically speaking, this can be

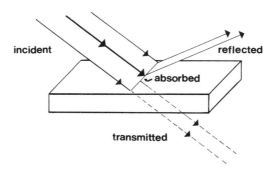

Figure 2.6. The alternative pathways of radiation falling on a surface of another medium.

written as the ratio of the object's emissivity, $W(\lambda)_0$, to that of a blackbody $W(\lambda)_b$:

$$\varepsilon(\lambda) = W(\lambda)_0 / W(\lambda)_b$$

If the emissivity of the blackbody is equal to 1, the emissivity of a graybody will vary between 0 and nearly 1.

Kirschoff's law also applies to black- and nonblack- (gray-) bodies. Therefore, at any wavelength and at any temperature, the spectral emissivity and spectral absorbance of an object are equal:

$$\varepsilon(\lambda) = \alpha(\lambda)$$

We can now define some radiation sources:

The *blackbody* has an emissivity equal to 1 at any wavelength:

$$\varepsilon(\lambda) = \varepsilon = 1$$

this also means that its absorption is 1 at any wavelength:

$$\alpha(\lambda) = \alpha = 1$$

Therefore, $\rho = \tau = 0$. A blackbody does not transmit any wavelength (i.e., it is *opaque*) and it does not reflect any wavelength.

A *graybody* may be transparent if $\tau(\lambda)$ is different from 0, or opaque if $\tau(\lambda) = 0$. In this case

$$\alpha(\lambda) + \rho(\lambda) = 1$$

If the body reflects almost totally the incident radiation, $\rho(\lambda)$ approaches 1. Thus

$$\alpha(\lambda) = \tau(\lambda) = 0$$

As $\varepsilon(\lambda) = \alpha(\lambda)$, this body does not emit. A perfect reflecting material such as a mirror or highly polished surface has an emission approaching 0. The emission of a graybody is expressed by the Stefan–Boltzmann law, modified as follows:

$$W(\lambda)_0 = \varepsilon(\lambda) W_b = \varepsilon(\lambda) \sigma T^4$$

2.5.1.3.3. Effect of Surface Characteristics. Most bodies are opaque to infrared radiation. Therefore, only the most superficial layer of the object contributes to the emission of infrared radiation toward the environment. For this reason, the value of T in the Stefan–Boltzmann equation is the temperature of the object's surface. For the same reason, emissivity $\varepsilon(\lambda)$ is directly related to the fine structure of the surface. For example, iron at 1646 K has an emissivity of 0.25 if its surface is polished, but 0.99 if it is rough.

2.5.1.3.4. Heat Exchange by Radiation. Consider a body B_1 which emits radiation toward objects constituting the environment B_2, e.g., the walls of a room. The walls partly absorb this radiation, but they also emit toward the body, which will absorb a fraction of the energy received. The net energy gained by B_1 corresponds to the difference between the energy it has emitted and that which it has absorbed; this is affected by many factors including all the thermal characteristics of the objects and the relative position of the body within its environment.

In practice, this can be described by the following equation:

$$Q_r = \sigma F_2^1 \left(T_1^4 - T_2^4 \right)$$

where F_2^1 is a factor embracing the geometric characteristics and various coefficients, such as the emissivity of the bodies. The problem is relatively simple when all temperatures are nearly equal. According to Wien's law, the main wavelengths will be almost similar. However, if the temperatures are different, the principal wavelengths are different and the coefficient F_2^1 is difficult to determine. Nevertheless, when the temperatures are very different, the problem does become simpler. The best example is that of an object subjected to the sun's radiation. Obviously, the radiation of the object toward the sun is negligible. Thus the energy transfer is only related to the energy emitted by the sun and to the absorption of the body for the wavelengths of solar radiation.

2.5.1.3.5. Secondary Emission. A body that absorbs heat, for instance from the sun, becomes warmer. It also emits energy through its own infrared radiation. If it becomes warmer, the main emission spectrum will vary. Consequently, solar energy absorbed by an object appears to be reemitted at modified wavelengths. This is the basis for the greenhouse effect: Since glass is transparent to solar radiation, this radiation is transmitted to the objects within the greenhouse. Inside the greenhouse, radiation is absorbed by the objects and they then emit radiation at their own temperature, at wavelengths different from those of solar radiation. However, glass is particularly opaque at such wavelengths. Therefore, radiation builds up inside the greenhouse and the temperature increases.

2.5.2. Heat Transfer with Change of State (Latent Heat Transfer)

In both man and animals, only one form of heat transfer with change of state occurs—that by water evaporation (vaporization). The transformation of water into water vapor requires energy. The amount of energy needed for 1 g of water is 0.685 kcal or 2.86 kJ. This value is called the *latent heat of vaporization* and symbolized by λ.

2.5.2.1. Transfer of Mass

For unit surface area, the rate of loss of water transformed from liquid into water vapor, $\dot{m}_{H_2O}$, is proportional to the difference of mass densities of water at the surface. It is also related to a coefficient of mass transfer, the equation being

$$\dot{m}_{H_2O} = h_D(\rho_{ws} - \rho_{wa})A_e$$

where h_D is the coefficient of mass transfer, and ρ_{ws} and ρ_{wa} are the mass densities of the liquid and the vapor respectively. (The concept of *mass density* is close to that of *concentration*.)

2.5.2.2. Transfer of Energy

The energy $\dot{E}$ lost by the liquid corresponds to the product of λ with each term of the previous equation:

$$\dot{E} = \lambda \dot{m}_{H_2O} = \lambda h_D(\rho_{ws} - \rho_{wa})A_e$$

It is possible to substitute for the mass densities partial pressures (P) of water vapor because

$$\rho = P/RT$$

(where R is the constant of a perfect gas) and because the temperature difference between both states is small. The previous equation then becomes

$$\dot{E} = h_D(P_{ws} - P_{wa})A_e/RT$$

where P_{ws} is the partial pressure of water vapor at the surface of the liquid (in practice the saturated vapor pressure at the temperature T of the water surface) and P_{wa} is the partial pressure of the water vapor in the ambient air. These pressures are measured in mm Hg or preferably in pascals. The driving force of this equation, i.e., ($P_{ws} - P_{wa}$), was recognized by Dalton in 1788. The ratio h_D/RT is the *coefficient of evaporation*, which is symbolized by h_e and expressed in kcal $\cdot$ h^{-1} $\cdot$ mm Hg^{-1} $\cdot$ m^{-2} or in W $\cdot$ mm Hg^{-1} $\cdot$ m^{-2}.

2.5.2.3. The Relationship between h_D and h_c

The analogy between evaporation and convection means that h_D can be calculated when h_c is known. At this point two new numbers should be considered, namely those of Sherwood (Sh) and of Schmidt (Sc). They are described as follows:

$$Sh = h_D L/D_w \qquad \text{and} \qquad Sc = \nu/D_w$$

where L is the geometric expression defining the exchange surface and D_w the coefficient of mass diffusivity of water vapor in air (ν has previously been defined as the kinematic viscosity).

In air, Prandtl's number and Schmidt's number are close to unity and consequently close to each other. This means that we also have

$$c_p \mu = k \qquad \text{and} \qquad \nu = D_w$$

Therefore Nu and Sh are equal. If each is divided by Re, we obtain

$$Nu/Re = Sh/Re$$

It is possible to rearrange this formula in order to obtain

$$h_D = (h_c/\rho_a c_{pa})(Pr/Sc)$$

which can be simplified into

$$h_D = h_c/\rho_a c_{pa}$$

where ρ_a and c_{pa} are the density and the specific heat of humid air respectively. This product varies from 0.25 to 0.32 when one passes from dry air at $0°C$ to saturated air at $50°C$. This relationship is known as the Lewis relation.

By the same reasoning, the coefficient of evaporation may be expressed as follows:

$$h_e = 2.21 h_c$$

if temperature is expressed in $°C$ and pressure in mm Hg. This relationship is very important in practice because it is easy to compute the value of h_e from that of h_c.

One therefore arrives at the following important conclusion: If we consider a surface of water A at temperature T, this temperature defines the saturated water vapor pressure P_{ws}^0. On the other hand, the water vapor pressure of the surrounding air is P_{wa} and the driving force is represented by the difference $(P_{ws}^0 - P_{wa})$. In terms of energy, the amount of energy $\dot{E}$ lost by the water during unit time is expressed by the following equation:

$$\dot{E} = 2.21 h_c (P_{ws}^0 - P_{wa}) A$$

The value $\dot{E}$ thus corresponds to the maximum quantity of heat that can be lost by vaporization from surface A. It is called the *maximal evaporating power of the ambience*. If the surface is not water itself but only a wet surface, the quantity of water lost into the environment is necessarily less than this maximum value. This problem will be discussed further in Chapter 7.

Temperature and Humidity Measurement

3.1. THE THERMOMETRIC SCALE

Temperature is not measured in a true sense; it can only be ranged on a scale. To establish a temperature scale, i.e., to assign a numerical value to one temperature, one must choose a body or body system, called a *thermometer*, which has at least one property that varies with temperature according to a relationship of the form $T = f(x)$, where T is the temperature of the thermometer at the value x of the chosen property.

The use of the thermometer is based on two experimental conditions:

1. When two isolated systems are brought into contact, their temperatures vary until they reach equilibrium.
2. When two distinct systems are in thermal equilibrium with a third system, the three are all at the same temperature.

All currently used scales are founded on two points. Although the first thermometer was conceived and built by Galileo in 1592, the first two-point systems were proposed at the beginning of the eighteenth century by Fahrenheit and Réaumur. Réaumur called the temperature of freezing water 0, and gave boiling water (at sea level) a value of 80 units. Fahrenheit proposed another scale with the value 0 for the temperature of freezing salt water and the value 100 for the temperature of a pyretic patient. However, as these points proved not to be constant, it became necessary to fix their respective values by reference to another scale. That scale, proposed by Celsius in 1742, is the one in current use and has provided a basis for defining thermo-dynamic units. Celsius used the same points as Réaumur, with the same value for freezing water, but with a value of 100 for boiling water. The degree centigrade, now officially known as the degree Celsius (°C), is the hundredth part of the difference between these two fixed values.

However, studies on the laws of thermodynamics have shown that temperature cannot be lowered beyond a certain value and this point has been called *absolute zero* or *thermodynamic zero*. In 1848 Lord Kelvin proposed a new scale, starting from this value, with the same interval in degrees Celsius. This unit is the *Kelvin* and it is symbolized by K.* In this scale, the temperature of freezing water is 273.15 K and that of the triple point of water (where the three forms —solid, liquid, and gas—are in equilibrium) is 273.16. (For this reason, there is theoretically a very small difference between the degree Celsius and the Kelvin, their ratio being 27315/27316.)

3.2. TEMPERATURE MEASUREMENT DEVICES

Many systems are used to measure temperature, but only those used in clinical medicine will be considered here.

3.2.1. Liquid Expansion Thermometer

The most commonly used device is the liquid thermometer. It is so well known that description is unnecessary. Its accuracy can be very high, but, in medical use, it is limited to measuring the tempera-ture of closed cavities. This is due to the fact that the whole bulb must be surrounded or immersed in the medium to be measured. It is not suitable for measuring skin temperature.

---

*Notice that the unit is the Kelvin (not the *degree* Kelvin), and that the symbol is K (not °K).

3.2.2. Thermocouples and Thermistors

3.2.2.1. Thermocouples

Thermoelectricity was discovered by Seebeck in 1821. The first attempt to use the thermoelectric effect for measuring the temperature of the human body was made in 1834 by A. C. Becquerel (the father of the discoverer of radioactivity). The principle of the *thermocouple* is based on a thermoelectric action and seems to be very simple. A circuit is made from two conductors assumed to be homogeneous but made of dissimilar materials. If the two junctions at the interface are at different temperatures, T_1 and T_2, there is a constant electric current in the loop. The voltage producing the current is determined by the values of these two temperatures, and by the material's composition (Figure 3.1). The direction of the

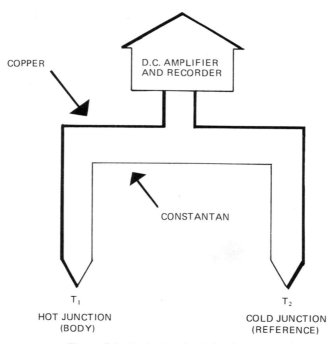

Figure 3.1. Basic circuit of the thermocouple.

current flow depends on which material is thermoelectrically positive, and whether T_1 is warmer or cooler than T_2.

In practice, the junction T_2 is maintained at a constant temperature; it is called the *cold junction* or *reference junction*. A simple but satisfactory method for obtaining a known constant for T_2 is to maintain the reference junction at 0°C, for example, by immersing it in a bath of melting crushed ice and water contained in a Dewar flask. The junction T_1 used for measuring the temperature at the chosen site is therefore the *hot junction*.

Unfortunately, the relationship between thermocouple voltage and temperature in nonlinear. It is usually specified in tables or graphs for each specific pair of materials. It is necessary to establish this relationship for every material used in this way (for example, see Table 3.1). The selection of pairs of thermocouple elements is based on a number of practical technical and biological criteria. For instance, the thermocouples used in medicine and biology must exhibit their best thermoelectric characteristics over the range of body temperatures. For this reason, only a few thermocouples are suitable for medicine, the most common materials being iron, copper, nickel, and an alloy of copper with 35–50% nickel that has been manufactured under numerous trade names, although it is now commonly called *constantan* (Ct). These materials are paired into Fe-Ni, Cu-Ct, and other types of thermocouples. Their sensitivity is small; for example, Cu-Ct produces a voltage change of 42 $\mu V/°C$ within the range of body temperatures (30–40°C) (Table 3.1). This low sensitivity was considered a disadvantage for some time, but now has been resolved by the use of more powerful DC amplifiers.

Thermocouples have important advantages, especially in experimental research. Many sizes of thermocouple wires are available and the thermosensors are easily constructed. Moreover, as only short lengths are needed in biomedical use, they are cheap to produce.

3.2.2.2. Thermal Clearance Techniques

A number of designs have been proposed for a probe aimed at thermal blood flow measurement through the skin, based on thermal conductivity. The probe commonly consists of an annular array of thermocouples with a heater, capable of raising local skin temperature by several degrees.

Table 3.1. Voltage Produced by a Cu-Ct (43% Ni) Thermocouple at Various Temperatures[a, b]

Temperature (°C)	Voltage (mV)									
	0	10	20	30	40	50	60	70	80	90
0	0.000	0.389	0.787	1.194	1.610	2.035	2.467	2.908	3.357	3.813
1	0.038	0.429	0.827	1.235	1.652	2.078	2.511	2.953	3.402	3.859
2	0.077	0.468	0.868	1.277	1.694	2.121	2.555	2.997	3.448	3.906
3	0.116	0.508	0.908	1.318	1.737	2.164	2.599	3.042	3.493	3.952
4	0.154	0.547	0.949	1.360	1.779	2.207	2.643	3.087	3.539	3.998
5	0.193	0.587	0.990	1.401	1.821	2.250	2.687	3.132	3.584	4.044
6	0.232	0.627	1.030	1.443	1.864	2.293	2.731	3.177	3.630	4.091
7	0.271	0.667	1.071	1.485	1.907	2.336	2.775	3.222	3.676	4.138
8	0.311	0.707	1.112	1.526	1.949	2.380	2.820	3.267	3.722	4.184
9	0.350	0.747	1.153	1.568	1.992	2.423	2.864	3.312	3.767	4.230
10	0.389	0.787	1.194	1.610	2.035	2.467	2.908	3.357	3.813	4.277
μV/°C	38.9	39.8	40.7	41.6	42.5	43.2	44.1	44.9	45.6	46.4

[a]The reference junction is at 0°C.
[b]*Example:* At 37°C, the voltage is 1.485 mV.

The system is heated after stabilizing with the skin, and the thermocouples measure the temperature rise of the skin until a new steady state is reached. The change in temperature will be a function of the thermal conductivity of living skin. However, with circulating blood present, there is an additive convective heat transport owing to the blood flow in the vessels in that area. Thermocouples should thus be situated at least 1 cm from the heater to take account of the skin temperature not directly influenced by the heater itself. The more capillary blood present in the area beneath the probe, the more heat is conducted away into the tissues, thus diminishing the rise in temperature caused by the heater.

Many questions relating to this technique remain unanswered. For example, the theory assumes that perfusion is homogeneous. Within a few millimeters of the surface this may be so, but this cannot always be assumed. The presence of large vessels can have a significant effect. Brown *et al.* have calculated that, with the type of probe they used, thermal flux due to conductivity in normal skin may be 38 mW, whereas that due to blood flow was only 7 mW. The question of the depth penetration of the heat delivered by the probe is also uncertain. Holti and Mitchell (1978) have performed experi-

Figure 3.2. Thermal clearance probe [courtesy of Ultrakust GmbH].

ments with layers of Perspex® sheets, since the conductivity of this material is close to that of bloodless skin. These experiments suggest that the penetration effect is on the order of a few millimeters.

These techniques are now drawing increasing attention and may be useful additive indices to the superficial perfusion of skin. However, they should be used cautiously and not claimed to be absolute measurements of blood flow.

A commercial thermal clearance probe is shown in Figure 3.2.

3.2.2.3. Thermistors

The *thermistor* is based on the observation that the electrical resistance of a semiconductor decreases as its temperature rises. The relationship between resistance R and temperature T is determined experimentally for each thermistor, and the temperature measurement thus becomes a measure of electrical resistance. Sensitivity is generally high with good accuracy.

Thermistors may be formed in a variety of shapes and sizes. They may be mounted on a probe and, according to the form of that probe, be put into all body cavities, e.g., vessels and tissues. They can also be placed on the skin and, if insulated from the ambient air, can usefully be used to estimate skin temperature. Their main problem is that they are "factory-made" and cannot be modified by the user. Furthermore, the relationship between electrical resistance and temperature, used to produce the calibration curve, is specific for each thermistor.

3.2.3. Liquid Crystals

Certain liquids are optically anisotropic and have other properties similar to those of crystals. Such substances are said to be in a *mesomorphic state*. In some cases, the optical properties are altered when the substance is placed under some form of stress (e.g., thermal, mechanical, or chemical). Several cholesterol compounds are mesomorphic crystals and change color with a change in temperature, making it possible to use this property as an indicator of temperature. A given compound will produce one color, which corresponds to one specific temperature.

In practice, *liquid crystal thermography* (LCT)* is the recording of the temperature distribution of a body surface by covering it with liquid crystals and monitoring the color changes. Liquid crystals can be placed directly on a surface by using a special paint; more commonly, they are mounted on a flexible plastic sheet, which is then pressed against the skin surface.

As with all contact methods for measuring biological surface temperature, liquid crystals, especially when used in sheets, can modify the thermal exchange properties of the surface, resulting in an alteration of the temperatures measured.

Liquid crystals are generally nonlinear, so in effect each color does not represent the same temperature interval. For example, blue may correspond to a range from 32.0° to 34°C, and red to a range from 35.0° to 35.6°C. Accurate quantification using these materials can therefore be very complex.

Figure 3.3[†] shows a liquid crystal thermogram of the palmar hand.

3.2.4. Radiation Thermometry

This method, which is increasingly being used in medical diagnosis, is based on the laws of radiation (see Chapter 2). All bodies emit electromagnetic radiation, and the total energy emitted in time units depends on the absolute temperature of the body. If this energy can be measured, it is theoretically possible to determine the temperature of the emitting body. In practice, certain factors must be taken into account before such a determination is possible, because (1) the emitted energy depends on the emissivity of the body surface and (2) the amount of emitted energy is very low.

Infrared radiometers are available for medical use. They have the advantage that skin temperature measurement of an area can be made without contact. The radiation emitted by the human body covers a large range of wavelengths. In 1934 Hardy showed that skin

---

*The term *contact thermography* is sometimes used to describe liquid crystal thermography. This should be avoided, because there are other methods for measuring surface temperature that also require contact between the detector and the surface.
†Figure 3.3 is included in the color insert that follows p. 42.

Figure 5.2. Thermogram of the scrotum. ⟶

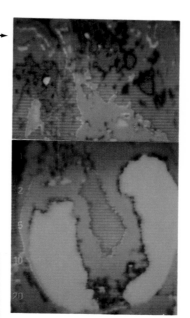

Figure 5.7. (A) Color thermogram of a male subject showing skin temperature distribution in a cool environment [courtesy of *Méditerranée Medicale*]. (B) Thermogram of the dorsal surface of a hand showing reactive hyperemia (paradoxical vasodilation) of the fingers. (C) Thermogram of a knee affected by rheumatoid synovitis. Note the hot area around the patella, in contrast to the cool pattern shown in (A) in the normal healthy subject.

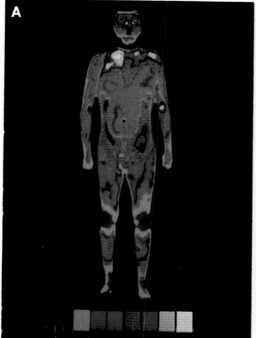

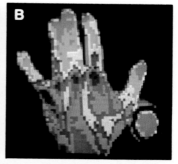

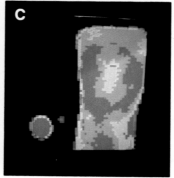

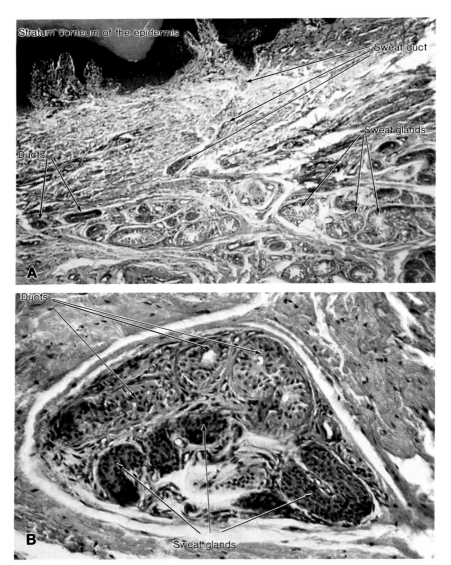

Figure 7.6. Cross-sectional diagrams of eccrine sweat glands under low (A) and high (B) magnification.

The two most common forms of picture presentation, monochrome or color, each have certain advantages and disadvantages. In the former, a continuous gray tone serves to highlight the blood vessels that lie near the skin surface. This can be of anatomic importance but has probably given some workers too limited a view of thermography. The *monochrome* picture is often enhanced by the electronic addition of isotherms. A defined temperature level and bandwidth can be added. This may mark the maximal or minimal temperatures in a scan; it may also be set over a specific feature and thus indicate its approximate temperature. If a temperature reference is incorporated in the scanner, or is included in the image, the isotherm values may be related to temperature (Figure 3.6). The image may also be indicated by more than one isotherm, but it is rarely possible to distinguish between more than two isotherms. In this case color techniques are of clear advantage.

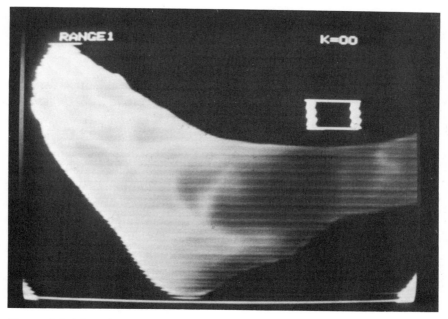

Figure 3.6. Monochrome thermogram showing the dilated vessels on the foot following chemical sympathectomy.

Color is becoming more common in medical imaging. It is advantageous to mark the maximum uptake of an isotope with a strong signal color, e.g., red. However, there can be dangers for the inexperienced thermographer, who can be diverted by the maximum temperature.

Color-coded thermograms must always be specified. Some processing systems will automatically set the upper level of the thermogram to maximum and spread all available colors over the temperature range of the thermogram. Others prefer to use a constant temperature color code. The selection of colors is also unstandardized in medical imaging. The most logical colors in thermography are those in the spectral order blue to red. However, the brain is affected by hue and vibrant strong colors attract more attention than pastel shades. If the number of colors available on the monitor screen is more than eight, varying shades of the same color are used. Some workers try to offset this by mixing the order of colors to give maximal contrast. Such color images are less recognizable and probably detract from the general acceptance of color thermography.

3.2.5.1. *Image Processing*

In addition to the use of isotherms, a range of facilities has been made available for thermography. The concept of a *thermographic index* from a standard window, or region of interest, has proved of value in the regular monitoring of certain diseases and their treatment. The simplest form is a deformable rectangle positioned over the site of interest on the analogue image. An integrated or mean temperature value is then displayed. Maximum, mean, and the difference between the two preselected windows can also be used. Calibration against an internal or external temperature source is required.

A wider selection of analytical methods is available when the scan is digitized, allowing interaction with a digital computer.

Picture storage may be achieved in analogue form by video-compatible systems on video tape. Digitized pictures can be stored on the whole range of magnetic systems available in computer technology, including tape, floppy disk, and hard disk. Storage has many advantages, e.g., allowing dynamic studies to be examined retrospectively. Picture comparison or subtraction may also be used to study change over a period of time. *Pattern recognition* techniques may help to solve the problem of the wide individual scatter in normal thermographic images. Further study of body temperature by thermal imaging methods is needed to clarify the criteria for specific clinical applications (Figure 3.7*).

Image processing techniques and the hardware necessary for their application are in continuous development. For this reason descriptions of these procedures and programs have not been included here. A number of articles, books, and conference proceedings are available that describe the methods. Anliker and Friedli (1976) and Ring (1979) are examples of applications of computerized thermography in breast cancer studies and rheumatology respectively. General coverage of techniques can be found in Onoe *et al.* (1980) and Dereniak (1981). A list of simple definitions related to image processing is given in "Thermographic Terminology," published by the European Association of Thermology in 1978.

3.2.5.2. Temperature Reference Source

Some form of *external temperature reference* is important for good use of infrared thermography. This is essentially a heated surface with emissivity characteristics close to those of the human body. The temperature control of the heater must at least equal if not exceed the sensitivity of the imaging system. Commercially available inverted cones and hollow tubes used as laboratory "blackbodies" can be too bulky for clinical use. In many instances it is desirable to include the reference in the field of view. This means that its bulk should be minimal, yet present a large enough surface area for calibration at distances up to 2 m. The commercial answer is usually a compromise with a flat plate within a protective cowl.

The reference source should be checked under varying ambient conditions, which may affect its temperature level. It should be located away from drafts. It can be used to check the linearity of the imaging system, for both spatial and temperature distribution. If a

---

*Figure 3.7 is included in the color insert that follows p. 42.

variable temperature source is used, the equilibration time for increasing or decreasing the temperature should be checked.

A large temperature source may be used as the basis for a *thermal phantom*. This is useful for quality control on the system, technique, and operator. It is particularly important where image processing techniques are used. A configuration of infrared attenuation filters, e.g., plastic sheets, can be used to create a standard or reference pattern. Thermal phantoms and thermal wedge sources are available commercially for industrial use.

However elaborate the equipment used, the quality and reliability of results are ultimately dependent on the technique and the operator. A strict protocol should be used in all thermographic procedures. Great care should be taken with the recorded image, especially if diagnostic importance is based on a photograph of the thermogram. This will be a very inefficient record of the patient or subject unless it is well documented, quoting such information as the temperature range of the scanner, level of reference temperature, position of the patient, and ambient conditions (see Section 3.2.7).

3.2.6. Microwave Thermography

The radiation spectrum of the skin and other objects includes not only infrared radiation but also a large range of other wavelengths, such as those in the optical and microwave bands, which extend from approximately 3×10^8 to 3×10^{11} Hz, corresponding to wavelengths of 1 m to 1 mm. In this field, however, the intensity of radiation is very small indeed. For example, the specific intensity of the microwave radiation of the human body is approximately 10^8 times less than the radiation at the infrared wavelength of 10 μm. On the other hand, the emissivity of human tissues within the range of microwaves is only 0.5. Nevertheless, *microwave thermography* is being developed in order to determine body temperature for two good reasons:

1. Although the emitted power is extremely small, detection of such power is within the capability of modern microwave radiometers, which were developed for radioastronomical observation.

2. Human tissues and other materials are partially "transparent" to microwaves. It is therefore possible to detect the microwave emission of subcutaneous tissues within the body and thus measure temperature without contact between detector and body.

The instrumentation consists of an antenna, which receives the radiation. This antenna is generally placed in contact with the skin (or the object's surface) to eliminate reflective loss at the tissue–air interface. It can, however, in certain cases, be placed at a short distance from the skin. A signal is then transmitted to the radiometer through multistage amplification. Calibration is usually made by comparison with a calibrated noise signal. Finally the output is fed to an analogue strip-chart recorder or a digital processor. Figure 3.8* shows a typical microwave imaging system and a representative scan. At present, these radiometers can detect intensity changes corresponding to emitted temperature changes of on the order of tenths of a °C.

3.2.7. Conditions for Thermal Imaging

It is clear that body temperature is subject to many influences. Thermal imaging and radiometry have broadened the use of skin temperature measurements in clinical medicine. But accurate and repeatable measurements are required, particularly for the regular follow-up of disease or its treatment. The patient should, therefore, be properly prepared for the examination in a suitable, stable environment.

The primary requirements for this kind of examination are:

1. Adequate insulation of walls, ceiling, and windows to assure temperature stability.
2. Control of air circulation and temperature.
3. Adequate space for cooling cubicles and necessary equipment.

The thermographic literature indicates that in northern Europe most workers have adopted 18–20°C as a standard ambient temperature. The reasoning is based on the optimal range between shivering

*Figure 3.8 is included in the color insert that follows p. 42.

and sweating levels. In fact this cool temperature environment produces good temperature contrast in the thermal image, particularly in the extremities. However the absence or presence of clothing on the trunk can greatly affect the skin temperature of the extremities.

Infrared thermography is sometimes used to evaluate vasospastic diseases, such as Raynaud's disease. In situations where the response to thermal stress is involved, optimal response will be obtained at higher ambient temperatures (see Section 7.1.3.1). Clinical use of reactive hyperemia can be made in the hands. Quantitative thermography is an excellent means of studying this phenomenon. The distributions of temperature, hence temperature gradients across an area, are simultaneously recorded in a single thermogram.

Since ambient temperature is important in thermography, it is recommended that monitoring systems be a permanent part of the examination room. Modern thermocouple devices with large digital displays are ideal and can be easily seen at a distance. The probe should be shielded against direct drafts and not be sited near heat-producing equipment, such as thermal reference sources.

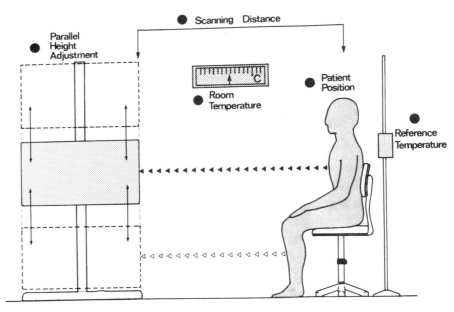

Figure 3.9. Factors in standardization of infrared thermography.

The requirements for the standardization of infrared thermography, particularly for examining locomotor diseases, have been outlined by a group report of the Anglo-Dutch Thermographic Society 1978 ("Thermography in Locomotor Diseases," 1979) (Figure 3.9). Although the *velocity of air* in the examination room is rarely quoted, air conditioning equipment is commonly used as an inexpensive means of controlling ambient temperature. A considerable range of air velocities may be obtained from such systems. The ideal conditions are probably provided by a laminar flow system. A compromise may be made by directing the air vents to a number of sites on the ceiling of the room. Jones (1975) has set out the optimal conditions for air flow in medical thermography. *Humidity* is also rarely quoted. Very high humidity levels will decrease cooling by the skin. The measurement of humidity is discussed in detail in Section 3.3. In practice, if adequate ventilation is provided, humidity is unlikely to present problems for thermal imaging or measurement.

3.2.8. The Patient

Standardized technique for thermography must include a strict protocol for patient preparation. Activity and food intake are variables that increase throughout the day. As a result temperature measurements appear more stable in the early part of the day. Many body parameters peak or acrophase around midday and skin temperature is frequently found to be higher at this time. For this reason absolute temperatures are more likely to be less variable before noon.

Patients who arrive at the thermography department from outside may have traveled in a cold or hot environment. It is essential that adequate time be given for rest in a stable (not cool) environment. This should be considered as a prerequisite to the usual preparation time of 15–20 min at 20°C. Smoking and drinking can affect skin temperature quite markedly in certain subjects.

Liaison with other hospital departments is necessary to obtain evidence of any medication being taken by the patient, which may influence temperature. Physiotherapy or sport activities must also be excluded in normal clinical investigations.

Finally, psychological factors must not be ignored. Reactions to the unknown, such as the first examination, vary greatly from one subject to another. Some stress can be prevented in anxious subjects through the maintenance of a careful and relaxed attitude by the staff. Full explanations of the procedures beforehand will facilitate the acceptance of those procedures. Since the anxious patient is not always readily identified, each patient should be treated with the same care and consideration.

3.3. MEASUREMENT OF AIR HUMIDITY

The humidity of air can be expressed in two ways: *absolute humidity* and *relative humidity*.

3.3.1. Absolute Humidity

Absolute humidity corresponds to the amount of water vapor contained in a unit volume of air. It is generally expressed in grams of water vapor per 1 m^3 of air. It can also be expressed in pressure units. In a given volume of air, water vapor, as does other gases (such as O_2 and N_2), produces a partial pressure related to its volume percentage and to the total pressure of the air. This expression is generally used in respiratory physiology. The humidity of alveolar air, for example, is approximately 47 mm Hg.

However, the amount of water vapor that can be contained in a given volume of air cannot exceed a maximal value. If we try to inject more water into this air sample, the excess water will not vaporize but will remain liquid. Expressed in pressure, this maximal value is called the *saturated water vapor pressure*, and the air is described as *saturated (water vapor)*.

Saturated vapor pressure depends on air temperature. The maximal value of humidity that an air sample can contain is easily determined providing the air temperature is known. The relationship between air temperature and saturated water vapor pressure (or maximal mass) is specified by diagrams or charts, such as that given as Table 3.2. At body temperature, for example, the saturated vapor pressure is 47 mm Hg. This value appears in the formulas for alveolar air for this reason.

Table 3.2. Saturated Vapor Pressure (P_w) and Mass of Water Vapor (m) as Related to the Temperature of an Air Sample at 760 mm Hg[a]

T (°C)	P_w (mm Hg)	m (g/liter air)	T (°C)	P_w (mm Hg)	m (g/liter air)
−10	2.16	2.38	26	25.27	24.46
−9	2.34	2.56	27	26.80	25.86
−8	2.52	2.76	28	28.42	27.33
−7	2.73	2.97	29	30.12	28.86
−6	2.94	3.19	30	31.89	30.47
−5	3.17	3.42	31	33.76	32.15
−4	3.42	3.68	32	35.72	33.92
−3	3.68	3.94	33	37.79	35.76
−2	3.96	4.23	34	39.95	37.69
−1	4.26	4.53	35	42.23	39.71
0	4.58	4.85	36	44.61	41.82
1	4.96	5.24	37	47.12	44.03
2	5.33	5.60	38	49.75	46.34
3	5.72	5.99	39	52.50	48.75
4	6.14	6.41	40	55.38	51.27
5	6.58	6.85	41	58.40	53.89
6	7.05	7.31	42	61.55	56.63
7	7.55	7.80	43	64.86	59.49
8	8.09	8.33	44	68.31	62.46
9	8.65	8.88	45	71.92	65.57
10	9.25	9.46	46	75.69	68.80
11	9.89	10.07	47	79.64	72.17
12	10.56	10.72	48	83.76	75.68
13	11.28	11.41	49	88.06	79.33
14	12.03	12.13	50	92.55	83.12
15	12.84	12.90	51	97.23	87.08
16	13.68	13.70	52	102.12	91.18
17	14.58	14.55	53	107.22	95.45
18	15.53	15.44	54	112.53	99.88
19	16.53	16.38	55	118.07	104.5
20	17.59	17.37	56	123.84	109.3
21	18.71	18.41	57	129.85	114.3
22	19.88	19.51	58	136.11	119.4
23	21.13	20.66	59	142.62	124.8
24	22.43	21.87	60	149.40	130.3
25	23.81	23.13			

[a]After Guerin (1952).

If we take a sample volume of air that is saturated and cool it, the saturated water vapor pressure decreases. This means that the amount of water vapor that could be contained in this air sample has also been reduced. Excess water vapor condenses, a process that explains the formation of dew and clouds. This process is used as a method for measuring air humidity and is therefore called the *dew-point method*. The humidity of air is determined by progressive cooling. The (theoretical) saturated water vapor pressure will decrease with temperature. When the pressure of the air sample approaches the water vapor pressure and continues to fall, condensation occurs. This in turn changes the optical properties of the air, which can be measured. Water vapor pressure is then determined from the dew-point temperature. For example, if we take an air sample at 25°C with an unknown humidity and progressively cool it, we find that the formation of dew just begins at 13°C. Table 3.2 shows that the value of saturated water vapor pressure at 13°C is 11.3 mm Hg. The humidity of this air sample is therefore 11.3 mm Hg (or 11.4 g/m^3 air).

Knowledge of the absolute humidity of air is important in the study of heat exchange between the body and its environment. In part, this exchange occurs by evaporation and depends on the absolute humidity of air.

3.3.2. Relative Humidity

The relative humidity of air is expressed as a percentage derived from the ratio of the actual vapor pressure of an air sample to the saturated vapor pressure of this air at a given temperature. For example, an air sample at 25°C may have an absolute humidity value of 11.3 mm Hg, and the corresponding saturated water vapor pressure is 23.8 mm Hg (see Table 3.2). Therefore, the relative humidity is 11.3/23.8 or 47.5%.

This form of expressing humidity is most commonly used because such simple devices as the hair hygrometer can be used to measure relative humidity directly. A more accurate device for measuring relative humidity is the *psychrometer*. This instrument consists of two identical thermometers, one of which is exposed and is known as the *dry bulb*. The other, which is covered by a layer of woven material that is always wet, is known as the *wet bulb*. They are placed in the same air sample together. The exposed bulb will indicate the actual air temperature (dry temperature). The second bulb, however, will indicate a lower temperature, owing to the cooling effect of the water vaporization. The level of this evaporation is inversely related to air humidity; the higher the absolute humidity, the lower the evaporation, and, consequently, the smaller the temperature difference between the wet and dry thermometers. There is a close

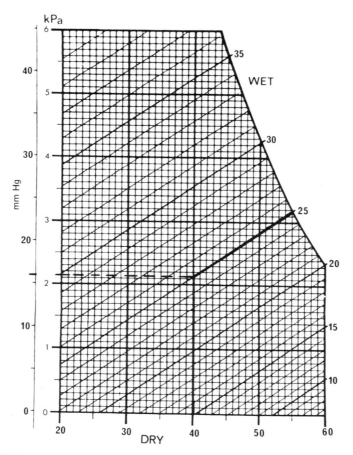

Figure 3.10. Measurement of air humidity by the psychrometric (wet and dry) method. *Example*: Wet bulb reads 25°C; dry bulb reads 40°C. Humidity is 16 mm Hg or 21.5 kPa.

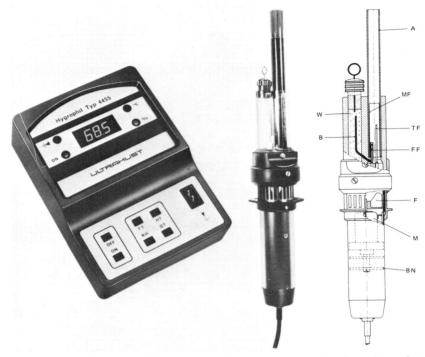

Figure 3.11. The Hygrophil™ instrument for measuring relative humidity: photo (left) and cross-section of temperature probe assembly (right). A, intake pipe; B, cotton wool stocking; BN, balancing network for probe; F, fan blade; M, fan motor; MF, metal foil; W, water reservoir; TF, temperature probe (dry); FF, temperature probe (wet) [courtesy of Ultrakust GmbH].

relationship between this temperature difference and the humidity of air. This relationship is generally presented as illustrated in Figure 3.10. A commercial temperature probe that utilizes the psychrometric principle is shown in Figure 3.11.

Although relative humidity is the most common form of humidity measurement, it should be used with care, especially in relation to thermal exchange between the body and its environment. Two values of relative humidity can be directly compared only if the air temperatures are the same. For example, if two air samples at 25°C have values of relative humidity of 25% and 50%, then the first sample is twice as dry as the second. However, if the air samples are not at the same temperature, the use of relative humidity can be misleading. If, for example, we consider one air sample at 0°C and another at 25°C, these samples may both have a relative humidity of 50%. The absolute humidity of the first air sample, i.e., the actual value of water vapor content, will be 2.3 mm Hg ($2.3 \ g/m^3$). However, the second air sample will be 11.8 mm Hg ($11.5 \ g/m^3$).

Therefore, when all systems are at the same temperature, humidity can be expressed in relative humidity. If the temperature is not the same, it should be expressed in absolute humidity. The latter condition generally applies to the human body in its environment, where the respective temperatures of body and environment are different. The heat exchange occurring by evaporation from the body to the air is therefore related to the difference between the absolute water vapor pressure of the body/skin and air. For this reason, the use of relative humidity in a medical and physiological context is of little consequence, and, strictly speaking, should be discontinued.

Man and His Environment

The fundamental problem of homeothermy is that the heat produced by the body as "physical waste" must be lost into the environment. If this heat cannot be totally or partially lost, the body temperature rises; on the contrary, if metabolic heat production is insufficient to compensate for the heat removed by the environment, body temperature falls.

Metabolic heat is produced by all tissues; however, the rate of production may vary from one organ to another. To simplify the study of heat transfer, one considers the body as composed of at least two parts: a *core*, which is composed mainly of the viscera, and a *shell*, which is the skin. Muscles may form a third layer between core and shell, but may also be included in the core (i.e., when the subject is resting or exercising at a low rate). The core appears to be that part of the body with temperature control, but the shell has an ambience-related temperature, and acts as a variable insulator for the body, facilitating temperature regulation at the core. In this simple concept, metabolic heat produced in a given unit of time by the core, $\dot{H}_m$, is transferred first to the skin. However, heat is simultaneously removed from the skin by the environment, and this heat loss in the same interval may be symbolized by $\dot{H}_e$. In a steady state, both forms of heat flow are equal.

The presence of clothing does not in principle modify this phenomenon but only creates another layer between skin and ambience. However, in a steady state, there is always equality between the

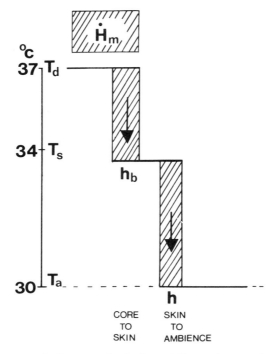

Figure 4.1. Heat transfer between the body and the environment can be compared to a two-step waterfall. The subject produces a certain amount of metabolic heat, $\dot{H}_m$, per unit of time (analogous to a certain mass of water per unit of time). Just as water falls from one level to another, so heat is transferred from the core (at T_d) to the skin (at T_s) (h_b is the transfer coefficient), owing to the temperature gradient. In a steady state, this amount of heat per unit of time is further transferred from the skin to the environment (at T_a), also owing to the temperature gradient. In the example shown in the figure, the temperature gradient is slightly higher than that existing between core and skin but the transfer coefficient h is lower. However, the areas of the three rectangles are the same, as they represent the same amount of heat transferred per unit of time.

three systems of heat flow, i.e., between the core and the skin, between the skin and the internal surface of the clothing (subclothing ambience), and, finally, between the external surface of the clothing and the environment. Such a system of transfer is represented in Figure 4.1 as a series of waterfalls. The differences between the temperatures (the driving force) are shown by the ordinate, the widths of the rectangles correspond to the heat transfer coefficients (which will be described later), and the areas of the rectangles represent the rate of transfer.

4.1. HEAT TRANSFER BETWEEN CORE AND SKIN

There are two ways in which heat is transferred between the core and the skin: tissue conduction and blood convection.

4.1.1. Tissue Conduction

As already described, conduction occurs when two solids at different temperatures are in contact. In the body heat is conducted between cells in the different organs. The rate of heat transfer, $\dot{H}_k$, can be written

$$\dot{H}_k = h_k(T_d - \overline{T}_s)A$$

where h_k is the coefficient of heat transfer by conduction of the tissues, T_d and $\overline{T}_s$ are the core and the mean skin temperatures, respectively, and A is the surface area of the transfer.

The coefficient h_k is closely related to thermal conductivity, which is symbolized by k in physics. Thus the determination of H_k would imply the measurement of k. This is difficult in practice, because this measurement implies first the suppression of blood circulation. One modality calls for the study to be done *in vitro* on excised tissue. Another one consists of temporarily suppressing the blood circulation of a limb, e.g., by means of a tourniquet. The general protocol is then to give a known amount of heat to the tissue and to determine how this heat is removed progressively from the tissue. This process is related to the *"thermal inertia"* of the structure, which is described by the expression $k\rho c$, where k is the thermal conductivity, ρ the specific mass (density), and c the specific heat. Since this product is directly determined by experiment, the conductive properties of the tissue are expressed by $k\rho c$, and not simply by the conductivity. Table 4.1 reports some values found in the literature.

Table 4.1. Values of Thermal Inertia of Some Human Tissues

Tissue	Thermal inertia ($J^2 \cdot cm^{-4} \cdot °C^{-2} \cdot s^{-1}$)	Comment
Skin		
In vitro	157×10^{-4}	Dependent on moisture content
Tourniquet-occluded *in vivo*	$96-131 \times 10^{-4}$	Dependent on extent of vasodilation
Fat	$38-56 \times 10^{-4}$	Temperature-dependent
Bone	$77-118 \times 10^{-4}$	Temperature-dependent
Muscle	$98-197 \times 10^{-4}$	Temperature-dependent
Forearm		
In vivo (segment)	253×10^{-4}	
Heel		
In vivo	122×10^{-4}	
Steel (for comparison)	1.57×10^{-4}	

4.1.2. Blood Convection

When blood flows through a tissue or an organ, it brings both fuel and oxygen and removes wastes. One such waste is the heat produced by the metabolism of that tissue or organ. Blood therefore gains heat, which is transferred by the circulation to the skin, where it is lost into the environment. (The respiratory tract is also a site where blood heat can be lost. However, its role is insignificant in man, with the exception of the neonate.) The term *convection* is generally applied to this process, although it is not correct physically speaking.

The heat convection rate, $\dot{H}_b$, depends on the temperature difference between core and skin and on the flow of blood arriving at the skin per unit time, $\dot{Q}_s$, and can be expressed by the following equation:

$$\dot{H}_b = \dot{Q}_s \rho_b c_b (T_d + \overline{T}_s) A$$

where ρ_b is the density of the blood and c_b is the specific heat of the blood. Some authors have slightly modified this equation by replacing T_d with the temperature of arterial blood (T_{art}) and $\overline{T}_s$ with the temperature of venous blood (T_{ven}). However, because of the anatomical patterns of skin vessels, particularly veins, it is not possible to assume that $\overline{T}_s$ corresponds to the venous blood temperature: Superficial veins may conduct blood from deep muscle, whereas blood that has passed through the skin may be transferred by deep veins.

The rate of total heat transfer from core to periphery, which must, in a steady state, equal the metabolic heat production rate, $\dot{H}_m$, is therefore expressed as the sum of $\dot{H}_k$ and $\dot{H}_b$:

$$\dot{H}_m = \dot{H}_k + \dot{H}_b = h_k(T_d - \overline{T}_s)A + \dot{Q}_s\rho_b c_b(T_d - \overline{T}_s)A$$

However, the role of blood convection is considerably more important than that of tissue conduction. In practice, the total amount of heat is measured and the expression of $\dot{H}_m$ can be simplified into the following:

$$\dot{H}_m = h_b(T_d - \overline{T}_s)A$$

where h_b appears as the sum of h_k and $\overline{Q}\rho_b c_b$. Although this coefficient represents mainly a transfer by convection, it is often called *blood conductance*. This term is not correct and should be replaced by *blood convectance* or better *core-to-skin transfer coefficient*. The main characteristic of this coefficient is its variability. This is due to skin blood flow, $\dot{Q}_s$, which can be largely modified by the vasomotor activity of the skin vessels: Vasodilation increases $\dot{Q}_s$ but vasoconstriction decreases it. These modifications occur during thermal regulation, mainly against heat. They will be studied in detail in Chapter 6.

Since the rise in h_b is generally due to an increase in ambient temperature (Figure 4.2), there is an inverse relationship between h_b and the temperature difference $T_d - \overline{T}_s$ or T_b: The higher the ambient temperature and h_b, the lower T_b. It is very difficult in practice to define the lowest T_b compatible with a heat transfer sufficient to ensure a homeothermic state. The temperature of 0.5°C appears to be the lowest value. It is very important to note that it is an absolute condition that mean skin temperature be lower than core temperature. This is because heat produced by the core (and muscle) must always flow toward the skin before it is lost to the environment.

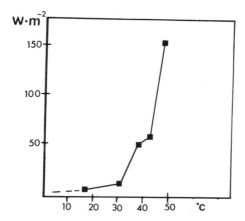

Figure 4.2. Heat produced by blood (h_b) at different body temperatures.

Indeed, it can be shown that skin temperature can be higher than T_d, e.g., when a part of the body is heated by an infrared lamp. T_s may increase locally to values higher than T_d, up to the onset of pain. However, this increase involves only a localized part of the skin and mean skin temperature remains lower than T_d.

In extreme conditions, mean skin temperature can itself be higher than T_d, but this can occur only in transient states, because the heat will flow from skin to core. This condition does not provide thermal homeostasis of the body. The simultaneous observation of an inverse difference ($\overline{T}_s > T_d$) and a thermal steady state is sometimes suggested but is physically impossible.

4.2. INSULATION

Chapter 2 described the concept of a thermal insulator as coming directly from that of a thermal conductor. The principle follows that of electricity, where the concept of resistance is related to that of conductance. Therefore, skin and subcutaneous tissues (mainly adipose tissue), may be considered as resistance to heat transfer between core and environment. This *thermal impedance* or *thermal insulation* is generally symbolized by I_s and is equivalent to $1/h_s$.

Clothing acts as a thermal buffer between the environment and man's body surface. In essence, it provides a new environment around the body. Not only does clothing itself act as an insulator, whose properties vary with the type, style, and fabric, but it invariably provides pockets and blankets of relatively still air underneath and within the fabric, providing additional insulation. Insulation reduces the rate of heat exchange by convection and conduction regardless of the direction of exchange, i.e., to or from the body surface. Moreover, clothing generally checks infrared radiation, producing a marked decrease in radiative heat transfer. The thermal impedance of clothing may be symbolized by I_{cl}. Its inverse, $h_{cl} = 1/I_{cl}$, is the heat transfer coefficient of clothing.

As in electricity, total thermal impedance of body skin and successive clothing layers is the sum of the different thermal impedances:

$$I_{total} = I_s + I_{cl_1} + I_{cl_2} + \cdots + I_{cl_n}$$

This makes the calculation of total thermal impedance relatively simple.

Clothing also acts on the thermal balance in another way. In a physically active man, it may add to his metabolism by providing a mass that must be moved through space, against air resistance. In addition, it may restrict body movement and produce unnatural movement.

4.3. HEAT TRANSFER BETWEEN SKIN AND ENVIRONMENT

This process follows the laws of thermokinetics described in Chapter 2.

Some heat transfer occurs within the respiratory tract, mainly by convection. The air flowing out from the respiratory tract is generally warmer than the inspired air. There is also an involuntary exchange that occurs by vaporization. Although these methods of heat transfer can in some animal species be quantitatively modified to deal with thermal regulation, their importance in the human is only related to ventilation and will be described further in Chapter 5. In this chapter, only heat transfer between skin and environment will be considered.

4.3.1. Heat Transfer by Conduction

Heat transfer by conduction occurs when part of the body is in close contact with a solid.

In a steady state, the net conductive heat flux, $\dot{K}$, depends on the temperature difference between the skin and the solid object, $T_s - T_{ob}$, and on the contact area, A_k, according to the following equation:

$$\dot{K} = h_k(T_s - T_{ob})A_k \qquad [\text{in W}]$$

where h_k represents the conductive heat transfer coefficient. It is dependent on the thermal characteristics of the skin, but, more particularly, on those of the object. We have seen that in a transient state (see Chapter 2) these characteristics are the thermal conductivity k, the density ρ, and the specific heat c. These three parameters are generally linked in the product kρc. The greater the product, the higher the capacity for heat transfer. (Table 4.2 shows the values of kρc of some common materials.) In practice, this explains why the sensation of the temperature of a given object depends not only on

Table 4.2. Values of Thermal Inertia of and Sensation of Heat
or Cold Given by Various Substances

		Temperature (°C)		
Material	Thermal inertia ($J^2 \cdot cm^{-4} \cdot °C^{-2} \cdot s^{-1}$)	Threshold of pain (cold)	Range of comfort	Threshold of pain (warmth)
Steel	1.57×10^{-2}	14	29–32	45
Concrete	3×10^{-2}	4	27–34	54
Rubber	3.8×10^{-3}	−12	24–35	67
Wood				
Oak	2.3×10^{-3}	−20	22–35	74
Pine	7.7×10^{-4}	−53	17–39	84
Cork	1.9×10^{-4}	−140	5–42	150

its actual temperature, but even more on its $k\rho c$ values. An iron object at 0°C gives a sensation of cold more than an identical object made of wood at the same temperature, because the iron removes more heat from the skin than does wood. Conversely, if the object is at a higher temperature than the skin, the greater the $k\rho c$, the more heat is transmitted to the skin, and, therefore, the greater the sensation of warmth that is experienced (Table 4.2).

Another parameter of the equation above is A_k, the surface area of contact. It is often said that heat transfer by conduction is of little importance in man. This is true when the subject is standing, because the surface area in contact is limited to the soles of the feet. It is also true when the subject is lying down, because the $k\rho c$ values of the bed structure are usually low, even though half the body surface is in contact with it. However, heat transfer by conduction may be very important. It is maximal when both A_k and $k\rho c$ are large. The classic example is when the subject is lying on cold ground. The amount of heat lost may be considerable, which explains the ease with which hypothermia can occur. This loss is more marked if the thermal regulation of the subject is disturbed, e.g., if the subject is wounded or intoxicated.

4.3.2. Heat Transfer by Convection

The body exchanges heat by convection with air or sometimes with water, when immersed. These exchanges, as seen previously, are related to the temperature difference between the skin and the air, $T_s - T_a$, and to the exchanging surface area of the body, A_c. They also depend on the convection coefficient, h_c. The equation for net convective is heat flux, $\dot{C}$, is as follows:

$$\dot{C} = h_c(\overline{T}_s - T_a)A_c \quad \text{[in W]}$$

The most common situation is when the air temperature is lower than the mean skin temperature. This results in a heat loss for the body. However, air temperature can be higher than skin temperature. In that case the exchange results in heat gain for the body.

On the other hand, the h_c coefficient depends on the thermal characteristics of the fluid, and also on its speed of movement relative to the subject. If air is static, convection is said to be *free*. When air is in motion, convection increases according to the relative speed of the air and is said to be *forced*. Therefore, the removal of heat from the body by the environment, when $T_a < \overline{T}_s$, increases with air movement. On the other hand, if $T_a > \overline{T}_s$, the increase in air movement increases heat stress.

4.3.2.1. *Determination of h_c*

The convection coefficient can be theoretically determined assuming that the body is a cylinder or is composed of cylindrical structures. On this assumption, Nishi and Gagge (1970) concluded that, in forced convection, the relationship between the Nusselt number and the Reynolds number would be as follows:

$$Nu = 0.33 Re^{0.55}$$

Since $Nu = h_c d/k$ (see Section 2.5.1.2.3), it is then possible to calculate h_c. This determination of h_c is interesting because it is possible to predict the variation in h_c when one of its constituent parameters varies, such as the variation of air density with altitude.

However, the most usual way of determining h_c is empirically, from the equation

$$h_c = \dot{C}/\left(\overline{T}_s - T_a\right)A_c$$

This kind of determination provides values for h_c that are only usable in conditions similar to those in which the values were determined.

Since h_c is related to the relative air movement, its expression generally includes the value of the air velocity, U. On the other hand, in free convection, U is equal to 0 by definition. Since in free convection, U is equal to 0, one must use different expressions for h_c, one without U (free convection) and another with U (forced convection). Colin and Houdas (1967a) have used an expression applicable to both free and forced convection. This expression is not physically valid, because the concept of forced convection excludes the concept of free convection. However, in practice, it is possible to use only one expression, since the convection is either free or forced.

The values of h_c currently found in the literature are presented in Table 4.3.

Table 4.3. Experimental Values for the Coefficient of Convection in Air (h_c) at Normal Atmospheric Pressure

Conditions (subject position/ air flow)	Relative air speed ($m \cdot s^{-1}$)	h_c ($W \cdot m^{-2} \cdot {}^{\circ}C^{-1}$)	Authors[a]
	0.15–0.30	$8.6U^{0.5}$ or $7.1U^{0.62}$	Nelson et al. (1947)
Standing: transverse air flow	0.50–5.00	$7.2U^{0.6}$	Mitchell et al. (1969)
	0.20–1.20	$2.7 + 7.4U^{0.76}$	Colin and Houdas (1967)
Lying on a mesh support: air flow parallel to the body	0.03–0.50	$10.6U^{0.5}$	Gagge (1937)
	0.05–0.10	$11.6U^{0.5}$	Winslow and Herrington (1979)
	0.2–1.20	$2.7 + 8.7U^{0.67}$	Colin and Houdas (1967a)
Lying on a bed: air flow parallel to the body	0.15–0.50	$7.3U^{0.5}$	Buettner (1934)
	0.15–0.20	$7.9U^{0.5}$	Aikas and Piironen (1963)
	Free convection	3.7	Winslow and Gagge (1941)
		4.1	Nishi and Gagge (1970)
Sitting on an ergometer bicycle	Convection caused by pedaling 50/min	4.8	Saltin et al. (1968)
	60/min	6.0	Nishi and Gagge (1970)

[a]Cited in Colin and Houdas (1967a) and Mitchell et al. (1969).

When an oversimplification is possible, a formula like the following, proposed by Kerslake (1972), may be used, but only for forced convection:

$$h_c = 8.3\sqrt{U}$$

In fact, the methodology that has led to all these equations can be challenged. An error is probably made on the measurement of mean skin temperature. For this reason, there is no valid criterion according to which a more simplified formula may be judged to be less valuable. On the other hand, the use of the square root of the air velocity makes the calculation easier. A simplified formula like that proposed by Kerslake leads to results very similar to those obtained by the other formulas.

A special case occurs when the air pressure is not at 760 mm Hg, either at high altitude or in hyperbaric conditions. The change in air density modifies the coefficient h_c in such a manner that the lower the air density, the lower is the value of h_c, and vice versa. The development of the Prandtl, Nusselt, Grashof, and Reynolds numbers shows that pressure variation affects h_c in the same manner that it affects U. This leads to the inclusion of the ratio "actual pressure in mm Hg/760 mm Hg" in the formula defining h_c:

$$h_c = 8.3\sqrt{U\,P/760}$$

Another special case is represented by modification of the gas composition of air as observed in the hyperbaric chambers used in underwater work. The partial replacement of nitrogen by other gases like helium produces a modification of the density and specific heat. The determination of h_c calls for the use of the adimensional numbers shown above (see Chapter 2).

4.3.2.2. Thermal Exchange in Water

The only possible pathway of heat exchange in water is by convection. The most significant characteristic of this exchange is that the convection coefficient in water is markedly higher than that in air—more than 25 times the coefficient in air. Therefore, for the same skin-to-environment temperature difference, the heat removal, i.e., the cooling power, is considerably more significant in water, easily producing hypothermia. As in air, one can distinguish free convection (when there is no relative displacement between body and water) from forced convection. Swimming is an example of forced convection. However, as shown by Boutelier (1979), the shivering of a resting subject in water produces only small movement and even this water is still. In this case, there is forced convection.

The convective coefficient increases with the speed of the relative displacement.

For determining the value of h_c in water, Rapp (1971) used a theoretical approach and proposed the following equation for free convection:

$$h_c = 0.848 Gr^{0.25}$$

for water at 22°C. However, since Prandtl's number is more affected by temperature in water than in air, this equation was only valid for the water temperature studied. If air temperature varies, Pr must be introduced. Using an exponent of 0.54 for Pr, Rapp calculated that h_c in still water was about 94 $W \cdot m^{-2} \cdot °C^{-1}$.

Experimental determinations of h_c lead to very different values, as shown in Table 4.4. There are some explanations for the significant discrepancies observed, including errors in the measurement of skin-to-water temperature difference, assumptions regarding the shape of the analytical models, and probable heat transfer by countercurrents in the limbs.

All authors who have been interested in these problems have also noticed the role of the skin surface in these transfer systems, and especially that of the face. In some experiments, the face is out of the water. If it is immersed, heat loss is enhanced.

Because of this high convective coefficient, the thermal neutrality of water is higher than that of air: 33–34°C for Boutelier (1979). When the subject is exercising, the temperature for passive equilibrium is lower, e.g., 26°C for a subject working at three times his resting metabolism.

There is one more factor that plays an important role in the thermocouple of the immersed body: the role of the subcutaneous fat

Table 4.4. Experimental Values of h_c for Free Convection in Water

Authors[a]	Experimental conditions	h_c (W · m^{-2} · °C^{-1})
Goldman et al. (1965)	Mannikin in vertical position	46
Boutelier (1979)	Nude subjects, heads not immersed, in supine position	
	Water at 33°C	44
	Water cool, subject shivering	61
Gee and Goldman (1973)	Nude subjects, measurements with flux meters, various water temperatures	
	35°C	38
	32°C	96
	28°C	208
	24°C	358
	20°C	537
Nadel et al. (1974)	Same technique as for Gee and Goldman (1973)	230

[a]Cited in Boutelier (1979).

layer. Fat does not modify the skin-to-water convective coefficient but it does produce increased thermal insulation between the body core and the skin.

When the water is warmer than the skin, heat flows from water to skin with the same high convective coefficient. The warming of the body is enhanced by the fact that the body has no other pathways for losing heat (no possibility of evaporation). In this situation, thermal homeostasis is not possible, and the limit of its ability to regulate body temperature is rapidly achieved.

4.3.3. Transfer by Radiation

As do all other surfaces, human skin emits electromagnetic radiation. According to its temperature range (≈ 305 K), the wavelength at which the maximum energy is emitted is 9.5 μm, corresponding to the far infrared. The energy flux $\dot{R}$ emitted by the skin can be described by Stefan–Boltzmann's law and is proportional to the fourth power of the skin temperature:

$$\dot{R} = \sigma\varepsilon_s T_s^4 \quad [T_s \text{ in K}]$$

where σ is the Stefan–Boltzmann constant (see Chapter 1) and ε_s is the radiant emissivity of the skin.

4.3.3.1. Emissivity of Skin

As for other surfaces, the emissivity of skin varies according to wavelength. In the infrared range, around 9.5 μm, the emissivity of human skin is close to 1 (that of a blackbody), whatever its color, as demonstrated by the first studies of Hardy (1934) and generally confirmed thereafter.

However, for other wavelengths, skin emissivity is not 1, as shown in Figure 4.3. In the range of visible light, i.e., between 0.4 μm and 0.8 μm, emissivity and absorption are low, but a difference appears according to skin color: White skin has an emissivity that is lower than that of black skin (Figure 4.4).

On the other hand, the range of wavelengths corresponding to skin radiation is not restricted to the visible and infrared. The radiometric spectrum of the body is vary large. Particular attention is

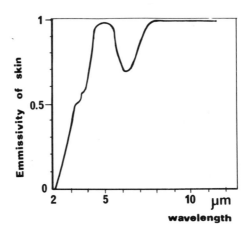

Figure 4.3. Emissivity of skin related to wavelength.

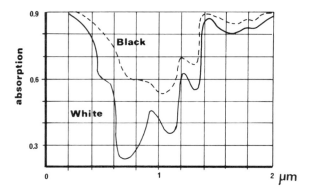

Figure 4.4. Absorption differences between white and black skin.

now given to radiation corresponding to frequencies of 1–10 GHz (microwave thermometry). However, the energy emitted at these wavelengths is extremely small (on the order of 10^{-14} W) and, in practice, the major part of the energy emitted by the body is emitted as infrared radiation.

4.3.3.2. Absorption by the Skin

As seen in Sections 2.5.1.3.3–4, the coefficient of absorption is measured by the same value as emissivity. This means that human skin absorbs completely infrared radiation at 9.5 μm (absorption coefficient close to 1). However, for other wavelengths, this coefficient may be less than 1, i.e., the skin does not absorb all the energy received.

4.3.3.3. Net Flux of Radiation

Skin emits energy toward environmental objects, and these objects emit toward the skin. The *net flux* is the difference between the energy received by and that emitted from the skin. It is difficult to define accurately the level of this net flux because of the numerous factors influencing these transfers, e.g., the geometric shapes, relative positions, temperatures, and emissivities of the objects. The emissivity closely depends on the surface characteristics of the radiant

Table 4.5. Emissivity and Absorption Coefficients
for Different Materials

Material	Emissivity and absorption coefficient
White marble	0.95
Oil paintings (all colors)	0.94
Anodized aluminum	0.94
Wood	0.93
Brick	0.93
Asphalt	0.93
Glass	0.90
White enamel	0.90
Copper, oxidized	0.87
Cast iron, oxidized	0.63
Aluminum, polished	0.04
Copper, polished	0.04

structure. A surface of polished aluminum emits little energy while the same surface at the same temperature, after being painted, emits more energy. Table 4.5 shows emissivity values for different materials.

In practice, the determination of the net flux of radiative transfer is empirical. However, two concepts introduced by Gagge *et al.* (1969) are very helpful in this determination. These concepts tend to simplify the problem of radiative heat transfer by taking account of the temperature of the radiant environmental objects.

4.3.3.3.1. Newtonian Radiant Flux. If an object is at an absolute temperature close to that of the body, its radiation is then in the infrared range. At these wavelengths the human skin acts like a blackbody: It emits and absorbs with a coefficient of 1. Under these conditions, it is possible to simplify the Stefan–Boltzmann law by "linearizing." It can then be expressed as

$$\overline{T}_s^4 - T_{ob}^4 = (\overline{T}_s - T_{ob})(\overline{T}_s^{-3} + \overline{T}_s^{-2}T_{ob} + \overline{T}_sT_{ob}^2 + T_{ob}^3)$$

The difference $\overline{T}_s - T_{ob}$ can be expressed either in Kelvin or in degrees Celsius. The second part of the right side of this equation may be symbolized h_r and called the *linear coefficient of radiative heat exchange*. As, by definition, $\overline{T}_s$ and T_{ob} are close together, this coefficient may be computed: It is generally obtained from tables or nomograms.

Since exchanges can be described by a linear relationship, they have been termed *newtonian radiant flux* (*NRF*). There is, however, an error in terming the relationship "newtonian." Newton established the equation for the temperature variation of an object placed in a cooler or warmer environment, but not the equation describing the transfer of heat, which was formulated by Fourier.

For an ambient temperature very close to that of the skin, h_r is on average $5.4 \text{ W} \cdot \text{m}^{-2} \cdot {}^{\circ}\text{C}^{-1}$.

On the other hand, the size of that part A_r of the total body surface area A that emits to and absorbs the radiations of environmental objects depends mainly on the position of the subject. Common values for the ratio A_r/A are given in Table 4.6.

4.3.3.3.2. Effective Radiant Flux. If the temperature of the environmental objects is very different from that of the skin, we can see that heat transfer occurs in only one direction.

Table 4.6. Estimation of the Ratio of Effective Radiating Surface Area of Body to Total Body Area in Different Positions

Position	Radiating surface/ total surface (A_r/A)	Authors[a]
	0.72	Bedford (1935)
	0.66	Guibert and Taylor (1952)
	0.70	Guibert and Taylor (1952)
	0.70	Fanger (1970)
	0.82	Bohnenkamp and Ernst (1931)
	0.82	Bedford (1935)
	0.78	Hardy and Dubois (1938)
	0.78	Guibert and Taylor (1952)
	0.84	Colin et al. (1966)

[a]Cited in Colin et al. (1970).

For example, in the case of the sun, energy is transferred only from the sun to the body. Therefore, the energy received by the body depends on the temperature of the radiant source and also the cutaneous absorption coefficient for the wavelengths of radiation, α_s, according to the Stefan–Boltzmann law:

$$R_{\text{received}} = \sigma \alpha_s T_{\text{ob}}^4 A_r$$

The measurement of emitted radiant energy can be made by a radiometer. The absorption coefficient of the skin is 1 for infrared radiation, but not for other wavelengths. Figure 4.2 shows the variation of σ_s according to wavelength.

Conversely, some objects can be at a very low absolute temperature. For example, a clear night sky may be at a temperature close to 0 K. In this case, the net flux of radiant energy corresponds practically to that emitted (and lost) by the skin.

$$R_{\text{emit}} = \sigma \overline{T}_s^4 A_r$$

In a heterogeneous environment in which some objects are at a temperature close to that of the skin (e.g., the walls of a room) and others are at a very different temperature (e.g., incandescent lamps, or, in industry, a mass of molten iron), radiative exchanges are complex, but they can be considered as if made of two parts—one corresponding to a temperature close to that of the skin, and described as the NRF, and another, which is algebraically added to this first one, which corresponds to the radiant sources at temperatures very different from that of the skin. This second has been called *effective radiant flux (ERF)*.

4.3.3.4. Exposure to the Sun

In the case of exposure to the sun, the radiant energy is known (Figure 4.5) but the problem is to determine that part of the body exposed and receiving this energy. Obviously this varies according to the respective positions of the subject and the sun. There are tables in the literature giving different values for skin surface heat exchange. However, it seems preferable to use the equation established

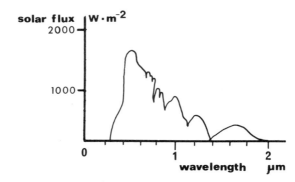

Figure 4.5. Solar radiant energy as a function of wavelength.

by Underwood and Ward (1966), who have assumed the body to be a cylinder. When related to the whole skin surface, the projective surface (or receiving surface) of this cylinder can be expressed by the following equation:

$$A_p = 0.043 \sin \theta + 2.997 \cos \theta \sqrt{0.02133} \cos^2 \zeta + 0.0091 \sin^2 \zeta$$

where θ is the elevation angle (angle formed by the axis of the sun and the horizon) and ζ is the azimuthal angle (angle formed by the sun and the frontal plane of the subject).

4.3.3.5. Combined Coefficient

When radiative heat transfer can be expressed as a linear relationship of the skin-to-object temperature difference (NRF), it is possible to combine the equation for convective heat transfer

$$\dot{C} = h_c(\overline{T}_s - T_a)A_c$$

with that for radiative heat transfer (NRF)

$$\dot{R} = h_r(\overline{T}_s - T_{ob})A_r$$

Providing that the temperature of the air and that of the objects is the same ($T_{ob} = T_a$) and that the parts of the skin exchanging heat either by convection or by radiation are also the same ($A_r = A_c$), it is then possible to write

$$\dot{R} + \dot{C} = h(\overline{T}_s - T_a)A_c$$

The coefficient h, which is the sum of h_r and h_c, is called the *combined coefficient for convection and radiation*. Its value can therefore be computed from Table 4.4.

4.4. BASAL EVAPORATIVE HEAT LOSS

Even in the absence of regulation (sweating, hyperpnea, etc.), a continuous evaporative heat loss takes place via the skin and respiratory tract. This small loss of water vapor, which can be measured simply by weighing the body, was discovered by Sanctorius in the early seventeenth century. He called it *insensible perspiration*.

4.4.1. Insensible Perspiration through the Skin

Even when there is no sweat secretion human skin is continuously losing a small amount of water vapor. This loss is due to a diffusion of water vapor from the skin and is even observed in a corpse. This loss of water vapor also corresponds to a loss of thermal energy by the body.

Insensible perspiration is a physical phenomenon and is strongly dependent on the water vapor pressure difference between skin and ambient air, which is generally low. Table 4.7 shows the values commonly found in the literature for adult man.

However, for the same water vapor pressure difference the involuntary heat loss through the skin is higher in the newborn, owing to the relative thinness of its skin, and this loss can become very important in premature babies, whose skin layers are extremely thin and permeable. Up to 120 ml · kg^{-1} · day^{-1} of water can be lost through the skin—equivalent to a loss of approximately 290 kJ · kg^{-1} · day^{-1} (70 kcal · kg^{-1} · day^{-1}). This figure is in excess of approximately half the usually recommended intake of 500 kJ · kg^{-1} · day^{-1} (120 kcal · kg^{-1} · day^{-1}) for the premature infant.

Table 4.7. Insensible Perspiration $(g^{-2} \cdot h^{-1})$ for
Three Values of Skin-to-Ambient
Vapor Pressure Differences

	Partial water vapor pressure (mm Hg)		
Authors	30	20	10
Zollner and Thauer	12	9.5	6.5
Burch and Winsor	10		
Hale *et al.*	10.5	9.0	
Brebner *et al.*	13	9.5	5.5
Heerd and O'Hara	7.5	4.0	
Colin and Houdas	8.0	6.5	4.5

4.4.2. Insensible Perspiration through the Respiratory Tract

Expired air is always more humid than inspired air. There is, therefore, an evaporative loss of water and of heat. However, a small amount of convective heat transfer also takes place in the respiratory tract. For simplification these two processes can be considered together.

Convective and evaporative heat transfer in the respiratory tract are complex because of the alternating displacement of air during the respiratory cycle. This is an important factor in preconditioning the air before it enters the lungs. It is achieved in the following way.

Ambient air is generally cooler than the body. When passing along the mucosa during inspiration it gains heat by convection and water vapor by evaporation from the mucosa. The respiratory mucosa are slightly cooled in the process. This air conditioning occurs mainly in the upper part of the respiratory tract but is only totally achieved in the small bronchi. In the alveoli, air is in thermal equilibrium (37°C) and saturated at this temperature (47 mm Hg).

During the subsequent expiration hot and humid air coming from the alveoli passes along the mucosa, which were cooled at inspiration. Therefore, the temperature and the moisture of this air decrease slightly. When it arrives at the outlet, air is then cooler than it was in the lungs. It remains saturated but at a lower temperature. Its absolute humidity has also decreased.

Nevertheless, the net result of the heat exchange is that expired air is warmer and wetter than inspired air. Therefore, the body has lost water and also heat.

4.4.2.1. Exchange by Convection

The value $\dot{C}_r$ is related to the air volume ventilated per unit time (ventilation rate), $\dot{V}$, and to the temperature difference between expired and inspired air ($T_E - T_I$):

$$\dot{C}_r = \dot{V}\rho c(T_E - T_I) \qquad [\text{in W}]$$

where ρ is the density of the air (1 g = 0.880 liter) and c is its specific heat (0.24 cal $\cdot$ g^{-1} $\cdot$ °C^{-1}). Because of the alternate inspiratory warming and expiratory cooling of the same air mass the temperature of expired air does not really depend on the ventilated air volume. It depends mainly on the temperature of the inspired and, therefore, ambient air. This relationship is almost linear within the T_I range 0–50°C (Figure 4.6).

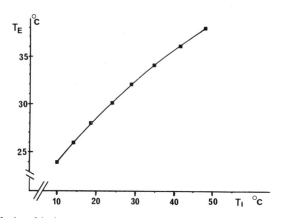

Figure 4.6. Relationship between expired air temperature T_E and inspired (ambient) air temperature T_I.

4.4.2.2. Exchange by Evaporation

Conversely, evaporative exchanges depend mainly on the ventilation rate. They depend also on the difference in water content between expired and inspired air, $C_E - C_I$. The amount of water vapor lost in one minute, $\dot{m}$, can be represented by the following expression:

$$\dot{m} = \dot{V}(C_E - C_I) \qquad [\text{in } g \cdot \text{min}^{-1}]$$

The formation of water vapor in the respiratory tract requires thermal energy. Thus, the amount of heat lost per unit time by this route is

$$\dot{E}_r = 2.45\dot{V}(C_E - C_I) \qquad [\text{in W}]$$

where 2.45 is the latent heat of vaporization of water (in $J \cdot g^{-1}$).

In practice, the measurement of this exchange is difficult. A nomogram was proposed by Houdas and Colin (1967), which is accurate but often too complex for a simple determination. However, since the heat exchanges via the respiratory tract, mainly the evaporative exchange, depend on the ventilation rate, and since this is itself dependent on metabolism, in practice a simplified formula can be used. Fanger (1970) proposed the following expression:

$$\dot{E}_r = 0.0027\dot{M}(44 - P_{wa})$$

In this formula, $\dot{M}$ is the metabolic energy production (in W), P_{wa} is the water vapor pressure of the ambient air (in mm Hg), and 44 is the accepted water vapor pressure of expired air (also in mm Hg).

In the newborn and young, the convective heat loss via the respiratory tract is also higher than in the adult, representing 1.4% of the total heat loss at 28°C but only 0.2% at 36°C ambient temperature. The evaporative heat loss via the respiratory tract is more important, representing 5.4% and 10.2% at 28°C and 36°C respectively with a relative humidity of 50%.

5

Temperature Distribution

Different areas within the body and the skin surface are not at the same temperature. This is because the temperature of a structure is dependent on the balance between heat production and heat loss.

At first, the balance between the skin and core temperatures appears to be quite different, the skin temperature being more dependent on environmental conditions. For these reasons, the skin and the underlying tissues are usually considered separately from the rest of the body. This division needs to be qualified but is interesting and clearly illustrates a key fact: The core can be truly homeothermic, since it is protected from the environment, but the skin, exposed to this environment, is, in a way, a poikilothermic structure.

The distribution of temperatures inside the core and over the skin will therefore by considered separately.

5.1. DEEP TEMPERATURE DISTRIBUTION

Each organ, and each part of that organ, has its own temperature. As already shown, its temperature depends on the determinants of its heat balance, which are numerous.

Heat production is quite different from one organ to another. Some, such as smooth muscle, have low metabolism. Others, such as the kidney, brain, or myocardium, have a high metabolic rate. The

Table 5.1. Oxygen Consumptions of Some Human Organs

Organ	Oxygen consumption $(ml \cdot min^{-1}$ per 100 g$)$
In vivo	
Left ventricle (resting subject)	7.8–10.3
Brain	3.4– 3.8
Kidney	About 10
Muscle	
Resting subject	0.15–0.20
Exercise of 180 W	About 11
In vitro	
Smooth muscle	About 1
Salivary glands	About 6
Liver	About 6
Stomach, colon, rectum	3–4
Lung	About 2

oxygen consumptions of various human organs are given in Table 5.1. However, heat production is not homogeneous within the same organ. Significant differences can occur from one part to another.

After its production by a tissue or organ, heat must be removed. This removal is performed by conduction from one tissue to another. Conduction depends on the temperature gradient between the tissues involved and on their thermal conductivity. Thermal exchange is mainly performed by blood convection. The heat loss of any organ is then particularly dependent on the blood passing through it, i.e., on the local vasomotor state. The temperature of the blood arriving at an organ is a major contribution to the thermal state of that organ.

Some representative temperatures within the core are listed in Table 5.2.

5.1.1. The Circulatory System

Temperature measurements within the circulatory system are performed with thermocouples or thermistors introduced by catheterization.

Table 5.2. Temperatures inside the Body Core[a]

Site	Low	High
Mouth	−0.45	−0.30
Esophagus	−0.30	−0.20
Stomach	−0.20	−0.10
Liver	−0.25	−0.05
Vagina	−0.05	+0.05
Brain (mainly hypothalamus)	−0.25	+0.05
Nasopharynx	−0.45	−0.40
Tympanic membrane		
Ambient air at +10°C	−0.40	−0.40
Ambient air at +40°C	+0.40	+0.40
External auditory meatus	−0.50	−0.10
Urine (at the meatus)	−0.15	−0.10
Cardiovascular system		
Pulmonary artery	−0.25	−0.15
Jugular vein		
Higher part	−0.05	0
Lower part	−0.25	−0.20
Vena cava	−0.30	−0.25
Renal vein	−0.25	−0.15
Coronary sinus	−0.05	+0.05
Right atrium	−0.25	−0.15
Right ventricle	−0.20	−0.15
Left ventricle	−0.25	−0.15

[a] Compared with rectal temperature in a resting subject, 36.85°C.

5.1.1.1. The Venous System

It is now obvious that venous blood is always warmer than arterial blood, except in the limbs, at rest. Therefore, in the inferior vena cava for example, the temperature gradually increases from the origin of the vein to its entry into the right atrium. The increase is stepwise, and results from the successive arrival of the different abdominal veins.

5.1.1.2. The Arterial System

From a thermal point view, the arterial system is not homogeneous. Certainly there is only a small gradient within the aorta and the great arterial vessels. In the limbs, however, when there is no significant muscular activity, venous blood is relatively cooler than arterial blood, because heat has been lost at the extremities. Because of the anatomical proximity between an artery and the corresponding vein(s), heat exchange by countercurrent occurs between them. Therefore the blood is gradually cooled in an artery but is progressively rewarmed in the vein(s).

5.1.1.3. The Heart

Mixed venous blood entering the right heart chambers includes warm blood coming from the viscera and a cooled portion coming from the skin. This is one reason why pulmonary artery blood temperature can be considered as a reasonable expression for mean body temperature.

Despite the fact that, in passing between them, the blood has crossed the pulmonary circulation, there is no significant thermal gradient between the right and left heart chambers. This observation is well demonstrated during cardiac catheterization. Actually, the alveolar air creates good thermal insulation. Moreover, the air has been completely warmed when it enters the alveoli, because the main heat exchange occurs when the air flows into the upper respiratory tract.

In humans, there is little data on myocardial temperature. Owing to the constant activity of the heart, it is logical to assume that its temperature is higher than that of the blood inside the heart chambers. Studies performed in animals confirm this assumption and show that there is a thermal gradient from the endocardium, which is warmer, to the cooler epicardium. The relatively high temperature of the myocardium is also confirmed by the classical observation, made during catheterization, that the temperature of the blood in the coronary sinus is higher than that of the blood in the right atrium.

5.1.2. The Respiratory Tract

Since there is an alternating air flow in the respiratory cycle, temperatures in the respiratory tract can vary from one point to another, or at the same site over time.

Air temperature increases from the entrance to the respiratory tract to the main bronchial tubes, where it is maximal in normal

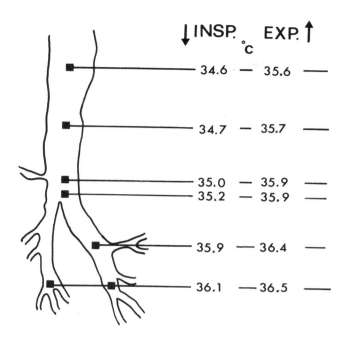

Figure 5.1. Temperature distribution within the respiratory tract (resting subject, lightly clothed, ambient air at 24°C, low ambient humidity).

ambient conditions ($T_a = 21°C$). Downstream, the air temperature is nearly the same as that of the pulmonary vessels. For this reason alveolar temperature is nearly constant and independent of ambient temperature variations (Figure 5.1).

Alveolar temperature cannot be directly measured. However there is a linear relationship between pulmonary capillary blood temperature and rectal temperature. If rectal temperature is taken to be the reference, pulmonary capillary temperature is 0.2°C lower and is then similar to heart chamber temperature.

5.1.3. The Digestive Tract and Abdominal Viscera

5.1.3.1. The Oral Cavity

Oral or, more exactly, sublingual temperature can only be measured reliably if the cavity is well closed. It is 0.3–0.5°C below rectal temperature. In most ambient conditions, oral temperature varies with rectal temperature. Conversely, if ambient conditions are unusual, oral temperature is then partially influenced by the environment. This probably occurs because of the proximity of the respiratory tract.

5.1.3.2. The Esophagus

Measurement of esophageal temperature is carried out by having the subject swallow a temperature probe. In the cervical region esophageal temperature is influenced by the air temperature in the upper respiratory tract. Conversely, in the lower esophagus it is close to cardiac temperature. This temperature is almost stable and is 0.2°C below the rectal temperature.

5.1.3.3. The Stomach

Gastric temperature was the first intraabdominal temperature to be measured (by Beaumont in 1833). It is relatively high, being 0.15°C below rectal temperature (Graf, 1959).

5.1.3.4. The Rectum

Rectal temperature is most often measured in clinical medicine or in experimental studies as representing the internal temperature of the body. In fact, using a mercury glass thermometer, what is probably more often measured is anal temperature. Part of the criticism against rectal temperature's being considered as the central temperature results from this simple failure. To obtain a true rectal temperature, the probe must be inserted to a depth of at least 8 cm.

When conditions are correct and the subject is at rest, rectal temperature is the highest in the body. This high value was first explained by the existence of bacterial fermentation. However, rectal temperature is not changed if the rectum is sterilized by antibiotics. This phenomenon can, however, be explained and seems to have a vascular origin. Blood flow is actually low in the pelvis, and a form of vascular isolation therefore occurs. Thermal insulation is important to this region, with high temperatures also occurring in other organs of the pelvis, such as the uterus and the vaginal rugae.

Rectal temperature is influenced by venous blood from the

limbs, the temperature of which can increase with muscular exercise. Nevertheless, it is important to note that rectal temperature exhibits a certain "inertia," even when temperature varies in other regions of the body.

5.1.3.5. The Abdominal Viscera

Because of its high metabolism, the *liver* has been considered the warmest organ in the body. It is in fact warm, but cooler than the rectum, if its temperature is measured at rest. Its temperature is 0.1°–0.2°C below rectal temperature and almost the same as stomach temperature. Hepatic temperature is not evenly distributed. Variations can be observed, depending on the digestive state.

Kidney temperature seems to be near that of the liver, but few data are available. Generally, measurements have been made in the renal veins.

The temperature of the pelvic organs, in particular the *female genital system*, is at the same level as rectal temperature. Temperature variations are of the same magnitude.

Fetal temperature *in utero* is 0.3–0.4°C higher that the temperature of the uterus owing to the relatively high heat production of the fetus.

In the human, as in most of the other species of mammals, the *male reproductive organs* are not situated in the abdomen. It is necessary to maintain the testes at a relatively low temperature to ensure normal fertility. External measurement of scrotal temperature is generally made by infrared thermography (Figure 5.2*). It is considered that this temperature may be representative of that of the testes themselves. However, microwave measurements will probably give more accurate information on testicular temperature.

5.1.4. The Nervous System

The metabolism of the nervous system is generally high. Jugular vein temperature is thus 0.2–0.3°C higher than superior vena cava temperature. The data available for *brain* temperature have mainly originated from animal studies. There is a noticeable difference between gray matter and white matter, the latter being colder.

*Figure 5.2 is included in the color insert that follows p. 42.

Since most of the thermoregulatory centers are located in the *hypothalamus*, this area has been well studied. As shown first by Hardy *et al.* (1964), the hypothalamic temperature is not homogeneous and increases from anterior to posterior. The anterior hypothalamic temperature is close to rectal temperature. Hypothalamic temperature exhibits some changes, especially when whole-body temperature is varying during muscular exercise or in hot environments. It has been observed that hypothalamic temperature is reduced during sleep, apart from the paradoxical phase, where a transient temperature increase appears in the preoptic area. These variations cannot be traced to regional metabolic changes. Blood flow variations could be involved. As in other areas within the core, the hypothalamus is cooled by arterial blood flow.

In the other areas of the central nervous system, temperature distribution has been less well studied, even in animals. The results show that the temperature of these areas is generally near the hypothalamic temperature.

5.1.5. The External Auditory Canal

This area, being easy to explore, has long been a popular site for body temperature measurement. The temperature of the cavity is, however, particularly dependent on ambient conditions, and therefore this temperature is not a true deep temperature.

The measurement is made on the tympanic membrane, and not in the auditory canal, as was done by Benzinger (1969) and coworkers. If the cavity is isolated from the environment, more accurate results are obtained. First, the measurement is easy to perform. In addition, this temperature must be close to the hypothalamic temperature because of the structure's anatomical proximity and partially shared vascularization. Studies made on animals show a good parallel between hypothalamic and tympanic temperature. In man, at rest, tympanic temperature is 0.05–0.25°C below rectal temperature.

5.1.6. Core Temperature and Its Significance

Obviously, there can be no *one* deep temperature. Several conditions are involved in the determination of deep temperature, and some are only regional. Nevertheless, the real problem is to know what is regulated and the respective importance of the core, peripheral sensors, and thermoregulatory centers (see Chapter 6). In any event, there is evidence from animal experiments that the integrity of some hypothalamic areas is necessary for thermoregulation. In addition, local change in hypothalamic temperature will induce appropriate thermoregulatory effector responses.

Because of its high thermal sensitivity, the central nervous system and especially the hypothalamus have been considered to be the ideal areas for deep temperature measurement. But since hypothalamic temperature is difficult to measure, more accessible sites have been sought that would have a comparable temperature range. In a steady state, tympanic temperature is lower than rectal, but it is close to hypothalamic temperature. Esophageal temperature is also lower than rectal. Oral temperature must not be considered as a correct deep temperature, as already shown, except under strictly controlled conditions.

When a sudden change appears in environmental temperature, the variation of tympanic temperature is more rapid than that of rectal temperature. Esophageal temperature change is parallel to that of the hypothalamus and is also more rapid. The difference can be explained by the "inertia" of rectal temperature. These facts have been confirmed during experiments utilizing extracorporeal circulation, as the esophageal temperature quickly varies with changes in infusion temperature but rectal temperature change is delayed.

In practice, when temperature is to be monitored during or after surgery, rectal temperature measurement is not suitable. The possible delay in diagnosis of a thermal abnormality could lead to an irreversible crisis. Rectal temperature can therefore be considered as a good approximation of the control temperature only when a subject is at rest, i.e., if his thermal balance is stable. Conversely, when variations appear and have to be controlled, tympanic or esophageal temperature measurement appears to be more appropriate.

5.2. PHYSIOLOGICAL CHANGES

Deep body temperature is never constant except in a steady state, which can only be obtained in experimental conditions. Physiologically, deep temperature is always changing to some extent and is dependent on many factors.

The first important factor is *age*. Significant differences in the central temperature and its regulation are found between the newborn infant and the adult. In addition, core temperature is subject to *cyclical* changes. Apart from such rhythmic variations, the main influence on core temperature variation is *muscular activity*.

5.2.1. Temperature in the Newborn

Before birth, there is a temperature gradient between the fetus and the mother (ΔT), with the fetal temperature normally higher. A decrease in this gradient would be pathological, and would occur as a result of subacute and prolonged stress, with ΔT decreasing as a function of the diminution of fetal heat production. Conversely, there is an increase in ΔT in acute fetal stress induced by asphyxia.

After birth, the colonic temperature of the fetus immediately drops. Both the thermal environment and newborn thermoregulatory responses induce this phenomenon. Thermogenesis does not switch on immediately after birth in man (see Chapter 7).

5.2.2. Cyclical Changes

Two types of cyclical temperature changes are very easy to observe: *circadian* variation and changes linked to the female *menstrual cycle*. In many other homeothermic species, a seasonal temperature variation can be seen. However, it is difficult to describe the same phenomenon in man, who generally lives in an artificial environment.

5.2.2.1. Circadian Rhythms

The existence of a circadian variation in body temperature has been well known since the eighteenth century. It has recently been shown that skin temperature also undergoes circadian change. These variations are not observed in the newborn but quickly appear when the thermoregulatory system comes to maturity. Temperature increases during the day, reaching its maximal level in the evening. The minimal value is observed at the end of the night (Figure 5.3). Several types of circadian variations have been described. They are

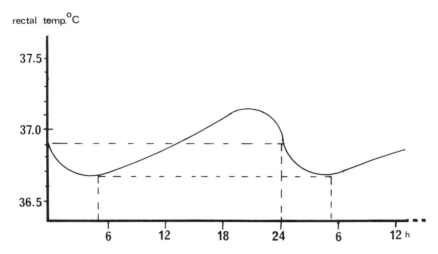

Figure 5.3. Diurnal variation of rectal temperature.

mainly dependent on the time of the maximal peak (*acrophase*), which can vary from one subject to another.

The relationship between central temperature variations and waking–sleeping alternation, i.e., light–darkness and also activity variation, is obvious. In fact, these connections are complex. Circadian high temperatures are certainly raised by muscular activity but are independent since circadian changes can still be observed in a totally resting subject. Studies performed on long-distance travellers have shown that very slow movement, e.g., by boat, is needed to induce a change in rhythm. It is always difficult to modify this period, and especially to reverse the phase of circadian variation. The variation is not due to the light–darkness alternation, since it is still present if light is constantly maintained at a given level. For this reason, therefore, time is not among the external determinants of circadian rhythm. In some experiments, the subject has been completely isolated from all external factors, such as light, time, and noise. It has then been observed that the period of central high temperature is maintained, even in experiments lasting as long as 6 months (Figure 5.4). The cycle then becomes slightly longer (24 hr 24 min during the first 2 months, 24 h 47 min during the last 2 months).

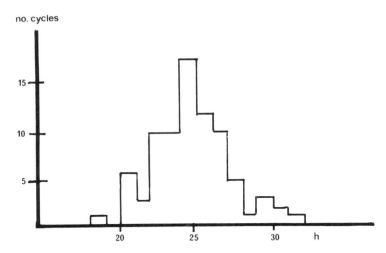

Figure 5.4. Histogram showing the distribution of the length of the circadian cycle during a 6-month free-running experiment. Note that the central value remains close to 24 h.

It is worth noting that the activity rhythms change considerably over the same time.

The period sequence for central temperature can therefore be considered as a specific property of the body, which can be modified by an external synchronization. The mechanism of this thermal cycle can probably be explained by a periodic change in heat transfer from the body to the environment. In fact skin temperature changes precede central temperature variations.

5.2.2.2. Temperature Variation during the Menstrual Cycle

Central temperature is well known to be a cyclical variant in the female. The mean temperature of the circadian cycle is lower during the postmenstrual period. It increases immediately after ovulation and is then maintained at a higher level through the premenstrual time. The rise is around 0.5°C but can vary in amplitude from one subject to another. Coincident with the cyclic change in deep body temperature there is an alteration in the subjective response to peripheral thermal stimuli. There is a marked increase in the cold threshold at the onset of menses with an equally marked decrease at the time of ovulation.

From a practical point of view, the menstrual change in deep body temperature is only obvious if the measurement is taken under the same conditions each day, to avoid interference by the circadian pattern.

5.3. OTHER CAUSES OF TEMPERATURE VARIATION

Apart from these cyclical variations, central temperature is dependent on many other physiological events, such as digestion and especially thermal stress.

5.3.1. Digestion

Normal digestion can increase central temperature. But this increase is very slight (0.1°–0.2°C). Nevertheless, certain food items significantly modify thermoregulatory conditions. Alcohol is one of these.

5.3.2. Thermal Stress

Variations owing to thermal stress are very important and worth further study. Thermal stress can be due to internal conditions such as metabolic rate changes, or external conditions, such as ambient temperature variations. The reactions of the body may be quite different at rest or while producing muscular activity, if its own metabolism is abnormally low or high.

5.3.2.1. Metabolic Rate

It is observed that the higher the metabolism, the higher the central temperature in steady state. There is therefore a relationship between T_d and the muscular activity. After a long period of exercise, when a new steady state has been reached, the central temperature can be as high as 40°C. It is therefore obvious that resting conditions are required to measure the basal value of T_d, which can then be a significant indication of the thermoregulatory state.

The connection between central temperature and metabolic rate is strictly physiological, except in extreme conditions that can be beyond the limits of the thermoregulatory system. During mild exercise, the body can maintain its central temperature at the normal value, using specific effectors. That it does not leads to a real theoretical problem (cf. Chapter 6).

5.3.2.2. External Thermal Load

Does deep temperature change as a function of ambient temperature? For a long time it was thought that they were quite independent.

Actually, the problem must be rephrased. Ambient temperature is not the only determinant of external thermal stress. The subject is

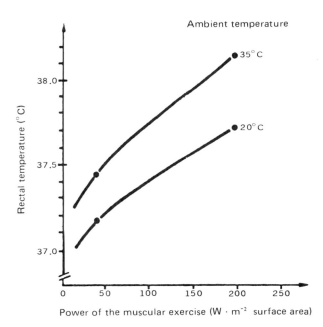

Figure 5.5. Relationship between rectal temperature and muscular exercise at two ambient temperatures.

often clothed and may be performing muscular activity. The connection between deep and ambient temperature is then of no significance if clothing or exercise is not precisely defined. It is therefore better to define a total thermal load as the sum of the rates of metabolic heat production and external thermal loading $[\dot{H}_m + (\pm\dot{H}_{ext})]$, rather than to use only the external loading rate $(\dot{H}_{ext})$, which is only related to T_d. All of the different factors involved are then considered, clothing included.

In addition, the relation can only be established in steady state conditions. When the whole thermal load is positive, the steady state can be reached rather easily. Conversely, when it is negative, i.e., in cold stress conditions, the steady state is much more difficult to obtain.

Apart from these problems, there is a clear relationship between central temperature and total thermal load whenever the subject is at rest or exercising (Figure 5.5). The relationship between deep temperature and metabolic rate is closer than that between deep temperature and external load.

5.4. SKIN TEMPERATURE DISTRIBUTION

The temperature distribution over an area of skin is dependent on the localized thermal balance. As the skin can be considered an interface between the body and the environment, its temperature is influenced by both internal and external conditions. When the external factors change and the skin is in a steady state, the thermal distribution may then be a function of the environmental conditions.

5.4.1. Skin Temperature Determinants

The internal and external factors that influence skin temperature can be described as follows.

5.4.1.1. Internal Factors

The skin itself produces a small amount of heat, which is evenly distributed over the total skin surface. Therefore, this is not a contributory factor in skin temperature gradients.

The skin is heated by tissue conduction, the heat being produced by underlying organs and tissues. Skin temperature is thus dependent on the thermal conductivity of the subcutaneous structures. Some areas, particularly those that are diseased, may have a higher metabolic rate and higher thermal conductivity. The skin overlying these areas is therefore warmer than the surrounding skin. This phenomenon is only observed if the pathological area is close to the skin or if it is extensive. A good illustration is breast cancer, which can induce skin temperature increase sufficient for detection by infrared or microwave thermography (see Figure 3.8B).

Blood convection is the main factor in warming the skin by transfer of heat from the core. Blood flow reaching the skin is therefore the chief internal factor in determining skin temperature. Any change in blood flow will induce a variation in skin temperature. The body is thus able to change its skin temperature, within certain limits, by modifying its vasomotor tone.

5.4.1.2. Ambient Factors

The ambient factors determining skin temperature are described in Chapter 4.

In practice, each area of the skin surface is exchanging heat with the environment. This partly depends on skin anatomy and on the local environment, which may differ from general ambient conditions. The location is important, since some areas are more exposed to the environment than others. This is obvious when one compares the forehead with the intermedial aspect of the thighs, where the surfaces are radiating to one another. Body position can also modify the thermal distribution of the skin by increasing heat exposure for some areas and decreasing it for others. There are therefore many variable factors involved in skin temperature production. Skin temperature must therefore be quoted under closely defined conditions.

5.4.2. Thermal Distribution in a Steady State

Thermal Neutrality. Thermal neutrality occurs when the subject is at rest and the thermoregulatory mechanisms are not required to maintain thermal balance. Body heat loss and basal body heat production are then equal. If the subject is at rest, this state occurs in

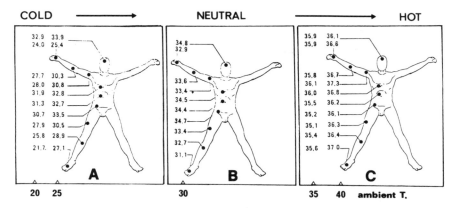

Figure 5.6. Skin temperature distribution in a steady state (B) and at colder (A) and warmer (C) temperatures.

an ambient temperature of 30°C with low humidity and air movement of less than 0.5 m · s⁻¹. Skin temperature distribution in a steady state at thermal neutrality is shown in Figure 5.6B. Infrared thermography (Figure 5.7*) is interesting because it reveals the local gradients, particularly obvious on the face and extremities.

When the environment is not in thermal neutrality (although in a steady state), skin temperature varies with ambient conditions. This variation differs with the particular skin area. The skin temperature distributions in a mildly cool (20 and 25°C) and also a warm (35 and 40°C) environment are described in Figures 5.6A and C. The extremities, especially the feet, show considerable variations in skin temperature. This also applies to the hand, but to a lesser degree. The forehead and trunk exhibit less significant variation, while the temperatures of the proximal limbs vary only moderately. It is therefore possible to determine the maximal skin temperature gradient for given ambient conditions. The colder the environment, the greater the range of skin temperature observed. At an ambient temperature of 35°C the temperature difference is minimal (Figure 5.8).

In clinical practice, it seems more useful to perform skin temperature measurements in a cold environment. Nevertheless, too low an ambient temperature should induce vasoconstriction and thus change skin temperature. It is not known if this change is the same in both

*Figure 5.7 is included in the color insert that follows p. 42.

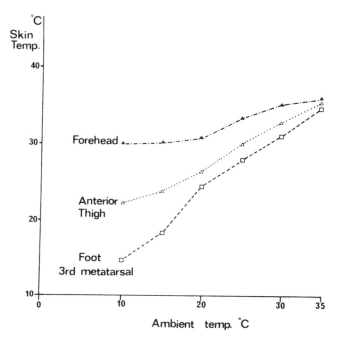

Figure 5.8. Relationship between skin temperature and ambient temperature in a steady state for three areas of the body.

pathological and normal skin. So, to study skin temperature gradients, the environment must be cool or neutral.

5.4.3. Thermal Distribution in Transient Conditions

The thermal environment of man is normally a changing one. When variation occurs and induces a change in skin temperature, the body is, from a thermal point of view, in a transient state. In order to study these transient conditions, the steady state before and after modification must be defined. The system is thus considered as a total entity, and the principal input and outputs are specified. The input is subjected to a known disturbance and the subsequent modification of the output is then studied.

When an abrupt disturbance appears in the thermal ambient, skin temperature varies but recovers to a stable level after a time.

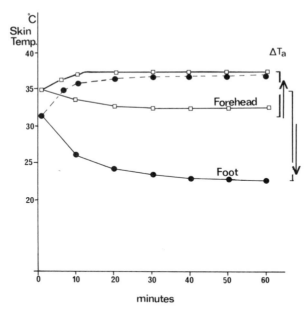

Figure 5.9. Skin temperature in an extremity, i.e., the foot, shows a greater temperature change resulting from an abrupt change in ambient temperature than does that of the forehead. Ambient temperature was increased from 30 to 45°C and decreased from 30 to 20°C.

The skin temperature curve, as a function of time, is an exponential. Skin temperature variation can then be defined by its *time constant*, i.e., the time during which 64% of the total variation occurs. This constant varies slightly from one skin area to another (Figure 5.9).

If the variations are low and occur in a neutral ambient, this constant is easily determined and will be approximately 5 min. At higher temperatures it increases and is more difficult to specify, due to sweating. The time constant in a warm ambient is about 10–15 min. However, in the cold this constant is shorter and will be approximately 8–10 min, unless the cold is extreme. In the latter situation the steady state cannot easily be reached, so the measurement of the time constant becomes very inaccurate.

In clinical practice, the concept of the time constant is an important one, especially when repeated thermographic measurements are performed on the same subject.

After a modification to the environment, the steady state ultimately reappears but at three times the time constant (using the analogy of the physicist's time constant). Therefore, for a subject undressing in a cool ambience, the steady state will be reached at best 20 min later.

5.5. MEAN SKIN TEMPERATURE

A single value representing mean skin temperature is very useful in helping to define the thermal balance more accurately. Unfortunately, it is only a theoretical notion, which might easily be criticized. "Mean skin temperature" is usually determined from a series of skin temperature measurements. However, the location of the measurement sites is quite arbitrary. In practice, the fewer the points used the better, but a sufficient number of points must be measured to obtain an accurate idea of mean skin temperature. The final choice from among the methods proposed by different authors becomes a compromise solution.

To relate skin temperature variation to underlying tissues, a coefficient can be assigned to each of the skin temperature measurements, depending on the site and the surface characteristics of the area. If the area is large and anatomically simple, e.g., the abdomen, the thermal gradients on the surface are low and one measurement with a high coefficient for the whole area can be used. However, the extremities are of complicated shape and possess high thermal gradients. For this reason, a greater number of measurements are required and the coefficients are lower.

The general formula for calculating mean skin temperature is

$$\overline{T}_s = c_1 T_{s_1} + c_2 T_{s_2} + \cdots + c_n T_{s_n}$$

where $c_1, c_2, \ldots, c_n$ are the coefficients affecting $T_{s_1}, T_{s_2}, \ldots, T_{s_n}$. Depending on the author, from 3 to 15 measurement points are determined in each case. These points and their respective coefficients are shown in Table 5.3.

The different methods have been compared by Mitchell *et al.* (1969), who have demonstrated that each method can be used for

Table 5.3. Sites Used for Calculating Mean Skin Temperature, with Weighting Coefficients

Site	Authors[a]											
	Winslow	Burton	Newburg and Spealman	Ramanathan	Palmer and Park	Hardy and Dubois	Neuroth	Omrec	Colin and Houdas	Houdas and Colin	Hardy and Dubois	Houdas et al. (1973b)
Forehead	1/15					0.07	0.12		0.06			
Cheek					0.14			0.1		0.2		0.07
Thorax	1/15	0.3	0.34	0.3	0.19		0.18	0.125	0.12	0.05		
Arms												
Posterior	1/15			0.3			0.08	0.07	0.08	0.05		
Medial										0.05		
Lateral (average)	1/15	0.4	0.15		0.11		0.05	0.07	0.06	0.05		0.19
Medial (average)												
Dorsal												
Head	1/15				0.05	0.05	0.04	0.06	0.05			
Abdomen	1/15					0.35	0.16		0.12	0.125		0.175
Thigh												
Anterior	1/15		0.33	0.2	0.32	0.19	0.18	0.125	0.19			
Medial	1/15							0.125		0.125		
Posterior	1/15											0.39
Leg												
Anterior	1/15	0.36·	0.18	0.2		0.13	0.11	0.15	0.13	0.075	0.065	
Posterior	1/15									0.075	0.065	
Upper foot	1/15					0.07	0.08	0.05	0.07		0.07	
Neck	1/15				0.19						0.0875	
Back												
Middle	1/15							0.125	0.12	0.2		
Lumbar	1/15										0.0875	0.175

[a]Cited in Mitchell *et al.* (1969).

most applications, even if only four measurement points are included. However, it seems more useful to measure both anterior and posterior points, as their environments can slightly differ. In addition, the underlying tissues have different effects on anterior and posterior temperatures.

However, as already shown, the criteria on which this method of determining $\overline{T}_s$ is based are subject to criticism. For example, the volume of a given limb segment is not an obvious criterion for determining that a high coefficient should be applied to that particular surface area.

Infrared thermography now makes it possible to determine the temperature at each point over the entire skin surface. Our own experiments have been carried out (1) to determine if any one of the proposed weighting formulas was better than the others, and, if not,

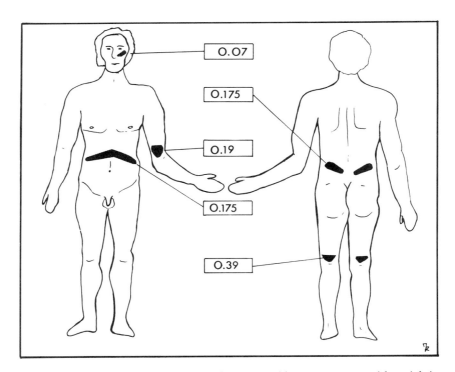

Figure 5.10. Proposed sites for calculating mean skin temperature, with weighting coefficients.

(2) to propose other points and coefficients founded on experimentally determined criteria. It is simple, using infrared thermography, to calculate the mean skin temperature of each cutaneous region. After applying the classical coefficients, it is possible to test if the calculated result agrees with the experimental value for mean temperature of the entire skin surface.

Different weighting formulas have been proposed for mean skin temperature. One that approaches the experimental results is obtained in the following way. Cutaneous points representative of each skin region are selected. The optimal coefficients are then calculated for any ambient, i.e., cold, neutral, or warm conditions (Figure 5.10). Since the surface area of the principal sites is reduced in cold conditions, the center of each area is measured to reduce inaccuracies.

5.6. MEAN BODY TEMPERATURE

Ideally, mean body temperature should be the mean of *all* measurements from both deep body and skin, but compiling these is a practical impossibility. It is therefore usual to define mean body temperature from mean skin temperature and one deep temperature, such as rectal or tympanic. Coefficients are usually given for $\overline{T}_s$ and T_d to express the relative importance of the shell and the core.

Mean body temperature can therefore be considered as

$$\overline{T} = x\overline{T}_d + (1 - x)\overline{T}_s$$

where x varies as a function of ambient temperature or, more precisely, of the thermal load. For most authors, x is around 0.7 (0.66) in the cold or at neutral temperature. At higher temperatures, peripheral vasodilation induces a change in the respective contributions of the shell and the core, increasing the value of core temperature. Therefore, x is even higher at around 0.8 (0.79) (Houdas and Sauvage, 1971).

The idea of a mean body temperature is of course theoretically interesting and useful. But, since its estimation is only approximate, it remains open to criticism.

6

Temperature Regulation

When a bird or mammal in a cold environment has the opportunity to escape to a warm one, generally it will do so. This reaction, which involves "conscious" adaptation, constitutes *thermoregulatory behavior*. If the animal cannot escape the cold environment, biological reactions, primarily an increase in thermogenesis, will occur in order to diminish the cold stress and prevent a fall in central temperature (*hypothermia*). These reactions are part of the physiological aspects of thermal regulation. They are, in general, purely involuntary. Behavioral and physiological thermoregulation also exist to prevent an increase in body core temperature (*hyperthermia*) when the animal is subjected to a warm stress. Ectothermic animals (*poikilotherms*) also exhibit behavioral thermoregulation but only the endothermic animals (*homeotherms*) have both behavioral and physiological thermoregulation.

The ability of homeotherms to maintain central temperature within the "normal" range solely by means of their physiological reactions varies greatly with species. However, whatever the intensity of thermal stress, when possible homeotherms first use their behavioral reactions in order to decrease the thermal strain on the body and, secondarily, their physiological mechanisms, to compensate for the insufficiency of the behavioral regulation and to achieve precise regulation of their central temperature. Physiological regulation therefore gives homeotherms the ability to live in different climates and to stabilize their body core temperature, allowing sustained daily activity. This results in the emancipation of the higher organisms from the effects of environmental changes.

However, the cost of physiological regulation is generally greater than that of behavioral regulation. The ability to prevent hyperthermia in a warm atmosphere very much depends on the water supply. This is because the most common way of losing heat is by evaporation of water through the respiratory tract and/or through the skin. Conversely, the maintenance of a high central temperature requires a high energy expenditure and in a cold environment energy consumption must be even higher. This may be a major increase, with 10–15 times the energy consumption at the basal level, and it chiefly supports the organism's production of more heat by shivering and/or nonshivering thermogenesis. The production of heat also depends directly on food intake.

In humans, the interplay between behavioral and physiological regulation is more complex for the following two reasons:

1. Because of the absence of fur, naked man has an extremely poor ability to resist cold stress but, conversely, he has a greater ability to resist warmth.
2. His behavioral control of both skin and core temperatures has been considerably increased by intelligence and extends to a variety of technological means such as clothing and air conditioning. These permit man to live in extreme environments, as encountered in polar regions or space flight.

In addition, thermal stress, however slight, is sensed to be uncomfortable by humans. Man uses his complex behavioral regulation not only to live in adverse climates but also to create a "comfortable" atmosphere around him and minimize or obviate the need for physiological regulation. Therefore, man's freedom with respect to climate is enormously enhanced but, paradoxically, his complex behavioral thermoregulation makes high energy demands.

6.1. THE "STEADY STATE"

When the body is said to be in thermal *steady state*, both superficial and deep temperatures are constant. This implies that the physical removal of heat by the ambiance exactly compensates for the heat produced by the body. Generally, a steady state is achieved when heat loss is equal to heat gain. If the sign $(+)$ is conventionally attributed to the gains and the sign $(-)$ to the losses, their algebraic sum will be equal to zero:

$$(\text{gains}) - (\text{losses}) = 0$$

Living bodies always produce heat. The main source of heat gain is metabolic. The letter $\dot{H}$ has been chosen to symbolize the rate of metabolic heat production. The subscript i ($\dot{H}_i$), or better m ($\dot{H}_m$), can be added to differentiate metabolic (internal) heat gain from possible heat gain from the environment. Effectively, the body can gain heat from the environment when the air and/or object temperatures are higher than skin temperature. However, this is less common, and generally the exchanges by conduction, convection, and radiation result in body heat loss. The expressions for rate of heat transfer by conduction, $\dot{K}$, convection, $\dot{C}$, and radiation, $\dot{R}$, may therefore be negative if they are heat losses, or less frequently, positive if they are heat gains. Moreover, the body can lose heat by evaporation. In neutral conditions, this loss is small. It takes the form of insensible perspiration from the skin and the respiratory tract. The rate of this evaporative heat loss can be termed *basal* and be symbolized by $\dot{E}_b$. It includes a small amount of heat transfer by convection in the respiratory tract. Obviously, it is always given a negative value.

Therefore, the fundamental equation of steady state thermal balance, which expresses equality between heat gain and heat loss, can be written as follows:

$$\dot{H}_m \pm \dot{K} \pm \dot{C} \pm \dot{R} - \dot{E}_b = 0$$

The symbol $\dot{M}$ is often used for the rate of total energy production of the body. To obtain the value for heat production, it is necessary to subtract the value for rate of production of mechanical energy, $\dot{W}$. In thermal problems, it seems more logical to use only thermal quantities.

For simplicity, it is possible to add algebraically the three modes of sensory heat transfer into one value, generally symbolized by $\dot{H}_e$. Effectively, this value is the heat flux removed from, or given to, the body by the environment. It can be considered as the external heat load or, rather, the external heat stress, regardless of its sign.

In a transient state, equality between heat gain and heat loss is not achieved. If heat losses are higher than heat gains, total body heat progressively falls. Therefore, the body is cooled and becomes hypothermic. If the heat losses are lower than the heat gains, total body heat rises and the body is progressively warmed to hyperthermia. This heat loss or gain by the body is termed *heat storage*, and symbolized by S (unit: J). When expressed as the difference between heat gain flux and the flow of heat loss, it corresponds to, and has the same dimensions as, heat flux. It can be then symbolized by $\dot{S}$ and is measured in watts (eventually $W \cdot m^{-2}$). Heat storage, and heat storage rate, are negative when the body is cooled and positive when the body is warmed.

The variation in body heat content results in a mean temperature variation of the body according to the following physical laws:

$$S = mc(\Delta T_b / \Delta t) \qquad \dot{S} = mc(d T_b / dt)$$

Where m and c are the mass and specific heat of the body, respectively.

Finally, the complete equation of the thermal body can be expressed as follows:

$$\dot{H}_i \pm \dot{H}_e - \dot{E}_b \pm \dot{S} = 0$$

It is very interesting to note that this equation can accurately describe thermal balance but does not give the actual value of the body temperature at which balance is achieved. It is therefore necessary to define the mean body or central temperature at which a given balance is achieved. The simplest steady state of thermal balance, at least theoretically, is that observed in "neutral conditions" when this balance is "passive."

6.2. THERMAL NEUTRALITY

By means of behavioral thermoregulation an organism tends to use physiological regulation to a lesser degree and always to be in neutral conditions. *Thermal neutrality* may be defined as the total condition allowing the organism, at rest, to be in a thermal steady state without active regulation, such as sweating or shivering. This state corresponds to the spontaneous equality between heat production, that is, the basal metabolic rate (internal conditions) and the heat removed by the environment (external conditions). This is achieved when the resting subject is placed in a homogeneous atmosphere at 28–30°C. Under these conditions, the body produces 50 $W \cdot m^{-2}$ and the environment removes exactly this value.

The application of the physical laws of thermokinetics, according to the various thermal characteristics of the body, especially the skin, shows that body temperature passively stabilizes at near 37°C for the core and about 33.5°C for the skin. A mannikin producing heat at a rate of 50 $W \cdot m^{-2}$ with a skin similar, thermically speaking, to that of man that is placed in a homogeneous ambiance at 28–30°C will spontaneously exhibit the same central and peripheral temperatures, without regulation. These temperatures appear to be the physical result of thermal balance. The level of internal temperature is very important but it is controlled by physiological, rather than physical, means.

Metabolic heat production is not constant throughout life. It is maximal during the growth phase of childhood and steadily decreases thereafter. This means that the parameters of thermal neutrality are also modified throughout life (Figure 6.1).

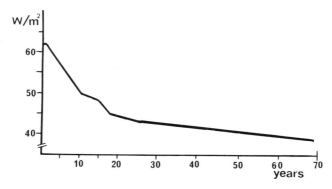

Figure 6.1. Variation of metabolic heat production with age. The high metabolism of childhood decreases with age.

In practice, the concept of thermal neutrality may be slightly enlarged. Man, being normally clothed, achieves thermal neutrality only if the environment temperature is lower than 28–30°C, depending on the value of the thermal insulation of his clothing. Moreover, the body is rarely totally at rest. Therefore, his metabolism is generally a little higher than $50 \ W \cdot m^{-2}$. For these reasons the external conditions for thermal neutrality correspond, in practice, to an ambient temperature of less than 28–30°C, i.e., to a temperature of 22–24°C.

A passive balance may be also created when a subject producing a lot of heat is placed in a cool ambiance. Then the removal of heat by the ambiance exactly compensates for the heat production. However the term *neutrality* cannot be used, by definition, since the subject is not in a basal state.

In fundamental studies of thermoregulatory function, it is always necessary to keep in mind that the conditions for defining true thermal neutrality constitute the basal or initial levels of all thermoregulatory mechanisms.

Thermal neutrality is therefore a very important concept; it may be considered as the reference thermal state of the body.

In conditions of thermal neutrality, the metabolic heat produced by the core is transferred to the skin according to the formula

$$\dot{H}_m = h_b(T_d - T_a) \qquad [\text{for } 1 \ m^2 \text{ of surface area}]$$

The environment removes heat at an equal rate from the skin. It is assumed that the basal evaporative heat loss is small and may be introduced by the symbol $\dot{H}_e$. (Note, that thermal sweating is absent by definition.) Heat loss to the environment may be expressed as

$$\dot{H}_e = h(T_s - T_a) \qquad [\text{for } 1 \ m^2 \text{ of surface area}]$$

Thus, we have

$$\dot{H}_m = \dot{H}_e \approx 50 \ W \cdot m^{-2}$$

It is possible to represent this heat transfer and thermal state using a scheme similar to that introduced in Chapter 4 (Figure 6.2). Heat is transferred at the same rate from core to skin and from skin

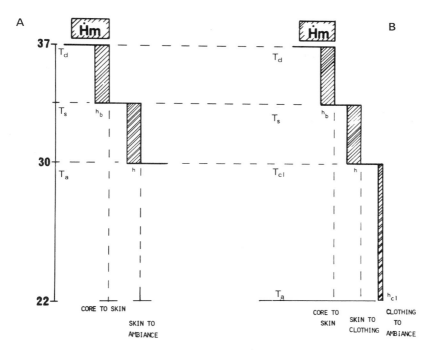

Figure 6.2. Conditions of thermal neutrality for a man, either nude (A) or clothed (B). In both cases man produces the same amount of heat in a given unit of time. Metabolic heat ($\dot{H}_m$) flows from core (at T_d) to periphery (at T_s) and from periphery to environment (at T_a). When the subject is nude, the environment must be at 30°C in order to ensure the full removal of $\dot{H}_m$. When the subject is clothed, a further stage for heat transfer is introduced: transfer from the immediate environment of the skin (created by clothing). $\dot{H}_m$ must be removed from this local environment by transfer through the clothing. Therefore, the final temperature will be governed by the thermal impedance of the clothing. In the example shown the final temperature is 22°C. For further explanation see the text (also Figure 4.1).

to environment. As illustrated, thermal neutrality of a clothed man implies that the environmental temperature has been decreased to take account of the insulation of the clothing. However, a clothed man has, by definition, the same thermal balance and the same thermal parameters ($\dot{H}_m$, T_d, T_s) as when naked. Figure 6.2B represents the steps of heat exchange in such a case, assuming that neutral ambient temperature is 22°C according to the thermal insulation of clothing.

6.3. CONDITIONS FOR PHYSIOLOGICAL REGULATION

If one or more of the thermal balance factors are varied, the equality between heat production and heat loss is broken down and thermal storage occurs. According to the imbalance, the temperature of the body tends to increase (hyperthermia) or decrease (hypothermia). To prevent too big a variation in temperature, the body triggers correcting (or regulating) mechanisms. However, it is very important to keep in mind that homeostasis implies that the body must necessarily lose its metabolic heat whatever the external conditions.

6.4. MECHANISMS AGAINST HYPOTHERMIA

The body tends (or would tend) to hypothermia (overcooling) when the removal of heat by the ambiance exceeds its own heat production. The net thermal balance is described in the following equation:

$$\dot{H}_m < \dot{H}_e + \dot{E}_b$$

Therefore, heat storage occurs corresponding to the difference between $\dot{H}_m$ and the sum of $\dot{H}_e$ and $\dot{E}_b$:

$$\dot{S} = \left(\dot{H}_e + \dot{E}_b \right) - \dot{H}_m$$

Such a disturbance can only be achieved by decreasing ambient temperature (T_a) (or one of the temperatures contributing to the ambient temperature, e.g., radiant temperature).

The first result of a fall in T_a is that skin temperature, which depends partly on T_a, decreases, although less than T_a. Deep body temperature also tends to decrease, but very slightly; this decrease can be neglected if the stress is low (Figure 6.3A). In the absence of regulation, the difference ΔT_a ($T_s - T_a$) is positive. As the combined coefficient h is unchangeable in a given ambiance, a positive ΔT_a implies an increase in the flow of heat removed from the skin by its surroundings (Figure 6.3B).

How can the body counteract this trend to hypothermia (behavioral regulation excluded)? The organism cannot act physiologically

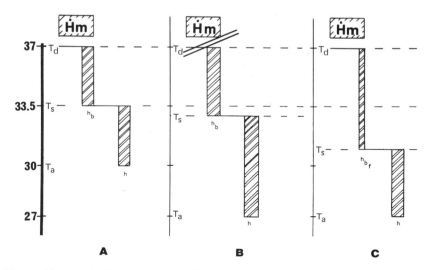

Figure 6.3. Regulation against a slight cold stress (vasomotor regulation) (ambient temperature: 27°C). (A) Neutral conditions. (B) Decrease of ambient temperature without regulation. Heat removal is increased and a steady state cannot be maintained. (C) Regulation is achieved by decreasing of h_b to the new value h_{b_f}. This induces a further decrease in T_s, but the heat removed by the ambiance is now $\dot{H}_m$ again.

on its environment, that is on h or T_a. It can, however, act on the only physiological factor involved in the exchanges, that is T_s. The decrease of T_s is achieved by lowering the heat support to the skin by a decreasing the core-to-skin heat transfer coefficient (h_b). The physiological reaction is a vasoconstriction of the skin vessels, which reduces cutaneous blood flow.

Therefore, the first response of the body to a fall in ambient temperature is to lower its skin temperature. When T_a is low, physiological regulation may produce a decrease in T_s so that the difference ΔT_a again equals its initial value. Therefore, the heat removed decreases to its initial value and a new balance is achieved without the need for further regulation (Figure 6.3C). The decreased skin temperature introduces only a mild sensation of cool discomfort. However, the possibilities for cutaneous vasoconstriction are slight and this vascular regulation is effective over only a very small temperature range (2–3°C above the neutral value). Yet in everyday

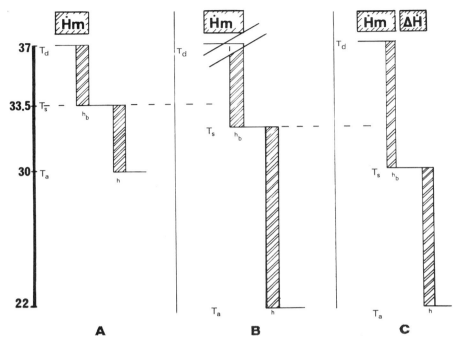

Figure 6.4. Regulation against a cold stress (ambient temperature: 22°C). (A) Neutral conditions. (B) Decrease of ambient temperature without regulation. Heat removal is markedly increased and a steady state cannot be maintained. (C) Regulation is effected partly by decreasing h_b, but this is not enough, and the body must increase its rate of metabolic heat production by a value $\Delta \dot{H}$ so that the sum $\dot{H}_m + \Delta \dot{H}$ compensates for the heat removal.

life, when the body is covered by clothing, this regulation may be very effective.

The body has only one additional means to prevent hypothermia: to increase its thermal production in such a proportion that the new heat production becomes equal to the heat loss (Figure 6.4). However, this response requires energy, whereas the energy demands of the vascular response are very low. (Actually, the body begins to increase its metabolic heat production before h_b has reached its lower value.)

Finally, if the vascular and thermogenetic responses cannot equalize heat production and heat removal, hypothermia becomes more and more marked, and death can occur.

6.5. CONDITIONS FOR HYPERTHERMIA AND MECHANISMS OF RESISTANCE

A tendency to hyperthermia (overheating) is realized when the metabolic heat production cannot totally be removed by the environment. The net equation describing this situation is as follows:

$$\dot{H}_m > \dot{H}_e + \dot{E}_b$$

Heat storage then occurs that corresponds to the difference between $\dot{H}_m$ and $(\dot{H}_e + \dot{E}_b)$:

$$\dot{S} = \dot{H}_m - (\dot{H}_e + \dot{E}_b)$$

In direct opposition to cold stress, which can be produced only by an increase in heat removal, heat stress may occur in two conditions: a decrease in heat loss by a rise in environmental temperature, and an increase in metabolic production to a level that exceeds heat loss. In fact, the two conditions may occur simultaneously and their respective heat stresses are then added.

6.5.1. Increase in Environmental Temperature

The increase in T_a leads first to an increase in T_s, but one that is smaller than the increase in T_a. T_d also rises, but, if the stress is slight, it is possible to assume that T_d remains constant. The increase in T_a leads first to a decrease of ΔT and reduced heat removal. The body cannot decrease its heat production, so heat storage takes place (Figure 6.5B).

The first and more rapid response of the body is to produce a further rise in T_s to increase heat removal to its initial value. This can be achieved by cutaneous vasodilation, which induces a rise in the skin blood flow and h_b. This regulatory mechanism may be enough to ensure a new steady state (Figure 6.5C).

However, the rise in ambient temperature may be greater and it may not be possible to remove a part of the metabolic heat production (Figure 6.6B). A very different mechanism is then used by the organism. In man, it is sweating. The evaporation of sweat has two

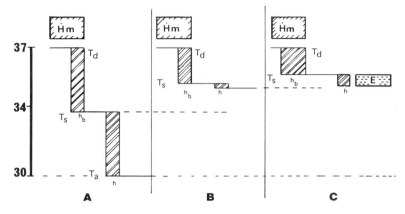

Figure 6.5. Regulation against a slight external heat stress (ambient temperature: 33.5°C). (A) Neutral conditions. (B) Increase in ambient temperature (T_a) without regulation: the heat removal decreases. There is a positive heat storage by the body. (C) Regulation is effected by increasing h_b, which results in an increase in skin temperature (T_s) and an increase in the skin-to-ambience temperature gradient.

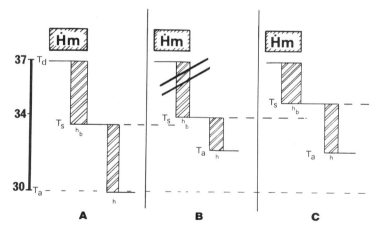

Figure 6.6. Regulation against a moderate external heat stress (ambient temperature: 35°C). (A) Neutral conditions. (B) Without regulation. Heat removal becomes almost zero whereas the metabolic heat cannot be totally transferred to the skin. (C) h_b is increased to ensure the transfer to the skin of the metabolic heat. However, this heat cannot be totally lost into the environment. Sweating and its evaporative heat loss compensate for the heat that was not been lost by sensible transfer.

results: (1) The skin is cooled in order to maintain a core-to-skin convection coefficient high enough to transfer the metabolic heat from core to shell. (2) The heat that is not removed by the driving force $(T_s - T_a)$ can be nevertheless removed by evaporation (Figure 6.6C), In fact, sweating occurs before the vascular reactions have attained their maximal value. The vascular reactions have low energy demands, whereas the sweating mechanism requires water. The total metabolic heat produced must therefore be lost by the evaporative pathway.

Finally, environmental temperature may be higher than that of the body. Without regulation, skin temperature would be higher than that of the core (Figure 6.7). Therefore, not only could metabolic heat not be transferred toward the skin, but an additional heat load would occur from the environment. Such a case would lead to rapid

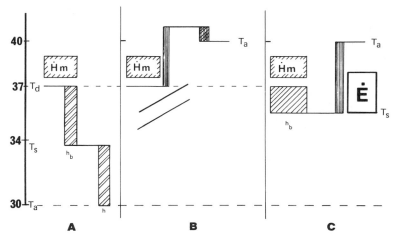

Figure 6.7. Regulation against a high external heat stress (ambient temperature: 40°C). (A) Neutral conditions: (B) Without regulation. Skin receiving heat from the core and the environment would increase in temperature because T_a is higher than T_d. Heat, therefore, could not flow from core to skin. (C) Evaporation of sweat first allows the skin temperature to decrease so that heat can flow from core to skin, but the organism must also compensate for the heat coming from the environment. This heat (represented as T_a) does not penetrate into the core (at T_d) but is "reflected" by the skin owing to the evaporation of sweat.

hyperthermia. The body has first to maintain a skin temperature that is lower than core temperature—an essential condition for achieving thermal homeostasis. This is achieved by the evaporation of sweat. Note that in this case vasodilation is not an efficient mechanism. On the other hand, evaporation must not only remove metabolic heat but also the heat flowing from the environment. A steady state may be obtained if evaporation exactly compensates for the sum of the metabolic and external loads (Figure 6.7C). If not, thermal homeostasis cannot be achieved.

A particular example is encountered when the environmental temperature is exactly equal to skin temperature. Without any regulation this would occur at a temperature slightly higher than 37°C. In fact, because of the cooling of the skin by the evaporative mechanism, on average this condition is achieved when $T_a = T_s = 35.5$°C. Since there is no sensory heat transfer, metabolic heat is lost only by the evaporative pathway.

6.5.2. Increase in Metabolic Heat Production

The first result of an increase in metabolic heat production, as in muscular exercise, would be unregulated, to produce a rise in T_d, followed by a smaller rise in T_s. Therefore ΔT_b would increase more than ΔT_a, which means that heat loss would increase less than heat production (Figure 6.8B).

The first body response is vasodilation, which increases the temperature of the skin and also h_b. Although ΔT_b is diminished, metabolic heat can be transferred to the skin. On the other hand, the rise in T_s increases the difference $T_s - T_a$; then heat loss can equal heat production.

If heat production is too great, so that vascular regulation is inefficient, the body must use evaporation. However, skin must be maintained at a temperature lower than T_d to ensure the transfer of increased metabolic heat to the skin. This means that the core-to-skin transfer coefficient must increase more and more, although there is a limit. Above this limit, metabolic heat cannot be transferred completely to the skin and hyperthermia cannot be counterbalanced (Figure 6.8C).

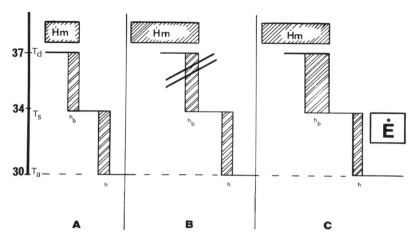

Figure 6.8. Regulation against an internal heat stress (increase in metabolic heat production). (A) Neutral conditions. (B) Without regulation. Heat cannot flow totally from core to skin. (C) Increase in h_b, allowing heat to be totally transferred to skin. At this level, sweat evaporation compensates for the heat that cannot be transferred by sensible pathways from skin to environment.

6.6. THE CONTROL NETWORK

The short-term control of thermal homeostasis is effected essentially by the *central nervous system (CNS)*. The CNS integrates the signals coming from the peripheral and central thermal receptors and, if thermal disturbance is detected, triggers the appropriate effector mechanisms in order to minimize or even cancel the disturbance (Figure 6.9).

Two stages are involved: First, structures that are sensitive to temperature and/or temperature variations, which are called *thermoreceptors* or, more generally, *thermosensors*, determine the level of thermosensitivity. Second, structures known as *thermoregulatory centers* integrate the signals coming from the thermoreceptors and trigger the appropriate effector mechanism(s).

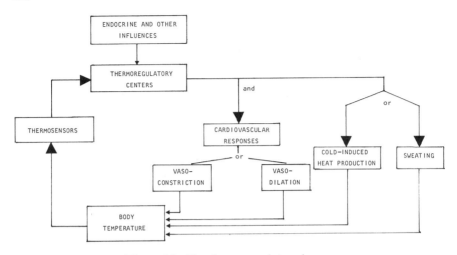

Figure 6.9. The thermoregulatory loop.

6.6.1. Thermosensitivity

Thermosensitivity is a property of almost all living cells. How-ever, in the study of thermal function this term is applied only to special cells that have a very low threshold for thermal stimuli and, for this reason, have marked thermosensitivity. They are thus able to inform the CNS of the temperature level of the body. These cells are called *thermoreceptors*. They are found in all species—even in ectothermic animals (poikilotherms)—which may utilize behavioral temperature regulation.

Thermoreceptors are located either in the skin or in the body core, especially in the CNS. They are termed *peripheral thermorecep-tors* and *central thermoreceptors*, respectively.

6.6.1.1. Peripheral Thermoreceptors

The thermoreceptors located within the skin are sensitive to their own temperature and send signals to the CNS. They provide conscious perception of ambient temperature and are able to stimu-late the thermal centers to bring about adequate behavioral and/or physiological responses. Electrophysiological data can be obtained by recording the impulse discharges propagated along the sensitive nerves or fibers when a skin region is thermically stimulated.

6.6.1.1.1. Responses of the Peripheral Thermoreceptor. Thermoreceptors, like other receptors in the body, exhibit two kinds of responses: static and dynamic.

a. Static Response. For a given stimulus, i.e., its own temperature, the response of a receptor, in a steady state, consists of a given number of impulses per second. There is therefore a fixed relationship between the temperature of the receptor and its discharge frequency.

This relationship is not linear. Two kinds of thermoreceptors have been distinguished, which have been categorized as *cold* and *warm*. Cold receptors increase their firing rate as the skin is cooled from about 38°C to 30°C and decrease their firing rate on further cooling. Warm receptors increase their rate of discharge as the skin is warmed from 25°C to 40°C after which there is a progressive decrease in firing rate up to 45°C. The relationship between cold and warm receptors is shown in Figure 6.10.

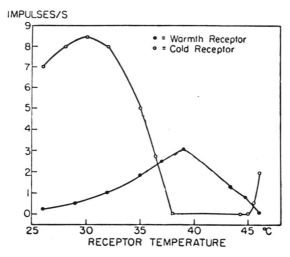

Figure 6.10. Relationship between the rate of discharge of warm and cold receptors and temperature in a steady state.

Owing to the particular shape of these curves there is a critical temperature at which the firing rate is maximal. However, this may be ambiguous, since two different temperatures applied to the same receptor can produce the same discharge frequency. Furthermore, cold receptors show a *paradoxical response*, corresponding to an impulse discharge at a high temperature, about 45°C.

Some fibers have been found that respond to mechanical stimulation and to changes in temperature in a manner similar to cold receptors. However, they differ from the true cold receptors in exhibiting a response characterized by a high firing rate at low temperatures that reduces as the temperature rises to 40°C. The question arises as to how the CNS differentiates the discharges related to mechanical or thermal stimuli. There is no answer at present and the significance of such receptors is not clear.

b. Dynamic Response. When temperature changes rapidly, the impulse frequency does not immediately reach its new level. There is an important difference between the two types of receptors. With increasing temperature, the warm receptors overshoot, their discharge frequency being transiently higher than the final level (Figure 6.11A). The converse is true for the cold receptors, which, for the same stimulus, respond by inhibiting the discharge signals (Figure 6.11B).

As shown by Iggo (1969), both static and dynamic responses are related "because the increase in discharge rate when temperature is changed is a constant proportion of the resting discharge rate." The integration of static and dynamic responses to both warm and cold receptors by the CNS minimizes the ambiguity of a single static response from one given receptor. This integrative function of the CNS provides an accurate assessment of skin temperature to the thermal centers. This integrative function is probably part of the overall function of these centers.

6.6.1.1.2. Distribution of Skin Receptors. The thermal sensitivity of the skin has been associated with the activity of certain skin structures, particularly Krause endings for cold receptors and Ruffini endings for warm receptors. Recent experimental findings, however, have demonstrated that thermal sensitivity is a function of free nerve

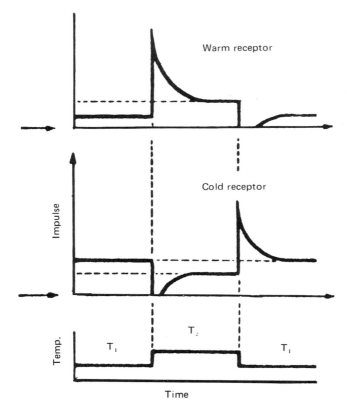

Figure 6.11. Dynamic response of the warm and cold skin receptors. A temperature increase from T_1 to T_2 causes a rapid rise in warm receptor discharge, returning to a new resting level. Conversely, the cold receptor is temporarily inhibited. When the temperature reverts from T_2 to T_1 the pattern of discharge is reversed.

endings, at least for the cold responses. These endings are located between the cells of the basal epidermal layer (Figure 6.12), at a depth of about 0.2 mm from the skin surface. The fibers conducting the "warm" signals are for group C, with a speed of discharge conduction of about $0.7 \ m \cdot s^{-1}$. The cold signals are transmitted either by nonmyelinated C fibers (speed of conduction: $0.7 \ m \cdot s^{-1}$) or by myelinated A fibers conducting at higher speeds ($6\text{–}11 \ m \cdot s^{-1}$).

The number of receptors per unit area of skin surface varies with the animal species and with the area of skin. The distribution is

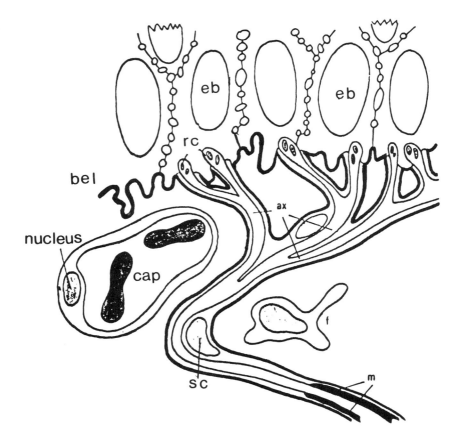

Figure 6.12. Schematic representation of the nerve endings observed on the site of a cold receptor in cat skin. cap, Blood capillary with cell nucleus and red cells; eb, epithelial basal cell; bel, basal epithelial layer; rc, receptive ending (with mitochondria); ax, axon; sc, Schwann cell; m, myelin.

generally higher on the face, including the ears and tongue, and, in the male, at the scrotum. However, regardless of regional location, the number of cold receptors is 10–15 times greater than that of warm receptors.

The receptive area for some fibers is a point 1–2 mm in diameter. However, other fibers may be affected by temperature changes in a receptive area of up to 1.7 cm².

Thermal signals may undergo a number of transformations during their passage through the spinal cord and CNS, including, for example, enhancement, inhibition, or summation.

Projections of skin thermoreceptors have been detected by recording from cells in the CNS, especially in the reticular structures, the thalamus, the sensory cortex, and the hypothalamus. The question that arises is whether the same skin receptor influences both the conscious cortical perception of temperature and the autonomic regulatory system. From what is know of the processes in the CNS, it is probable that the same receptor serves both functions.

6.6.1.2. Central Thermoreceptors

The term *central* is here used as the opposite of *peripheral*, and implies only that thermal receptors have been found in other parts of the body. They are most numerous, however, in the CNS, particularly in the anterior hypothalamus. These receptors respond by modifying their impulse discharges when their temperature is changed.

Changes in temperature in different parts of the CNS have been effected by different techniques, for instance by diathermy. Present investigations generally use a thermode. A *thermode* consists of a very small U-shaped tube that is sterotaxically introduced into the nervous structure to be studied. The tube is filled with water, the temperature of which is monitored, to induce the thermal disturbance, i.e., cooling or warming.

6.6.1.2.1. Hypothalamic Thermosensitive Neurons. The hypothalamus contains some neurons that modify their firing rate when a temperature change is made in their vicinity. However, the simple generation of a response does not mean that a neuron is fully thermosensitive. It may only receive and transmit a nervous signal related to this temperature variation.

On the other hand, there are in the hypothalamus some neurons that produce a thermal effector reaction when they receive electrical, but not thermal, stimulation. These neurons are considered part of the thermoregulatory centers.

a. Criteria for Thermosensitivity. Since all neurons show some sensitivity to temperature, by definition a *thermosensitive neuron* exhibits some specialized properties. The best evidence of this is founded on the study of the Q_{10} factor—the factor by which the metabolism of a given cell or region is multiplied when the temperature of this cell or region is increased by 10°C. The most common neurons have a Q_{10} of about 2. A Q_{10} greater than 2 denotes a particular sensitivity to temperature and is related to the property of thermoreception.

Hensel (1973) has pointed out that a true thermoreceptor must have both static and dynamic responses: This means that the sensitivity should not change when the temperature is kept constant. On the other hand, for the sensitivity to change with temperature, there should be a positive temperature coefficient for warm detectors and a negative one for cold receptors. True thermosensitive neurons have a low sensitivity to barbiturates.

b. True Thermoreceptors. In the hypothalamus, about 25% of the neurons exhibiting a thermal response can be considered true receptor neurons, or *primary thermoreceptors.* They are located histologically at the midline immediately supraventral to the anterior commissure (preoptic area) (Figure 6.13). They are stimulated by their own temperature and produce a nervous signal, i.e., an impulse discharge related to their stimulation. As already shown, these receptors are in general not very sensitive to pharmacological depressants. On the other hand, their discharges are not modified by the stimulation of the peripheral thermoreceptors.

Like the thermoreceptors of the skin, hypothalamic receptors exhibit a static and a phasic response. Different kinds of static responses occur, and may produce a linear increase of impulse frequency with temperature (Figure 6.14A), or a linear decrease with temperature (Figure 6.14B). The first is typical of the response of the "warm" thermoreceptors, the second, of the response of the "cold" receptors.

In the hypothalamus, the number of cold thermoreceptors appears to be very much lower than that of the warm receptors (ratio of 1 : 50).

However, this representation of the hypothalamic neurons involved in the detection of thermal signals is only schematic. These

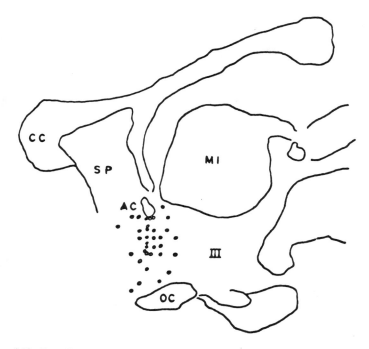

Figure 6.13. Localization of central thermosensors. ○, Warm sensors; •, cold sensors; CC, corpus callosum; SP, septum pellucidum; MI, thalamus (massa intermedia); AC, anterior commissure; OC, optic chiasma; III, third ventricle [from Hardy *et al.*, 1964].

neurons are only part of the nervous network of the hypothalamus and may also play a role in other functions such as the control of water and electrolyte metabolism.

c. Interneurons. Some hypothalamic neurons modify their firing rate when the temperature is locally modified. However, their discharge is not linearly related to temperature, as shown in Figures 6.14C, D. Moreover, their responses are inhibited more by pharmacological depressants, particularly the barbiturate group.

These neurons, representing 80% of the thermally responsive neurons, are not considered true receptors. However, it is generally accepted that they constitute a pathway for the signals issuing from the true receptors. They probably form the first level of integration of thermal signals.

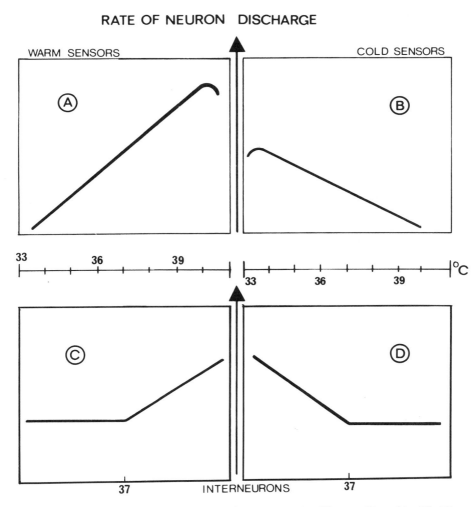

Figure 6.14. Response of some central neurons. (A) Warm; (B) cold; (C, D) interneurons.

6.6.1.2.2. Mesencephalic Thermal Receptors. Some neurons of the mesencephalic reticular formation exhibit a high thermal sensitivity. According to the criteria given above, 8% of these neurons can be considered as cold primary detectors and 79% as warm primary detectors, whereas the remaining 13% are interneurons. Histologically they have a wide distribution within the mesencephalic reticular formation dorsal to the red nucleus. At the present time, however, the role of the thermosensitive mesencephalic neurons is less well known than that of the thermosensitive neurons of the hypothalamus.

6.6.1.2.3. Thermal Receptors in the Spine. Many experiments show that cooling or warming the spinal cord in unanesthetized animals induces thermoregulatory responses. (Thauer, 1970; Simon, 1974). These responses are well adapted and may be as intense as those obtained by cooling or warming of the hypothalamic structures. It has been demonstrated that they do not depend on the stimulation of efferent fibers but are due to the stimulation of primary thermoreceptors. These spinal receptors exhibit static and dynamic responses, as do the other true receptors. The role of spinal thermoreceptors in relation to the hypothalamic receptors is not yet totally clear. It is, however, clear that the motorneurons of the spinal cord are themselves thermosensitive, for they increase their firing rate when cooled.

6.6.1.2.4. Other Central Thermoreceptors. In some animals, heating and cooling the abdomen has been seen to activate thermoregulatory response, suggesting the existence of thermosensitive structures in this region (Rawson and Quick, 1972). It has also been suggested that the walls of large veins and muscle fibers could contain thermosensitive cells. However, the existence of central thermoreceptors is not well established, and their possible significance in thermoregulation is not clear.

6.6.2. Thermoregulatory Centers

Knowledge of thermoregulatory centers and their patterns of activity has been gained by two techniques: localized destruction and electrophysiological techniques.

The first techniques used involved the local destruction of certain parts of the brain stem and/or sectioning of the brain stem itself. These first findings were compared to pathoanatomical observations made in humans. In this way it was possible to define the location of the principal nervous structures involved in thermal regulation. However, these techniques were neither accurate nor specific. For example, the persistence of a response after cutting the brain stem does not mean that the nervous structure triggering this response is situated below the section. It could be that the section had cancelled an inhibitory control coming from above.

More recent electrophysiological techniques have considerably added to the accurate localization of the nervous structures involved. These techniques allow us to measure the activity of several neurons or even single neurons by recording their action potentials with microelectrodes. The discharge of these neurons may be spontaneous, as during a thermal response to disturbance. It can also be experimentally produced by modifying the temperature of the thermal receptors, for example, with a thermode. Microelectrodes and thermodes may be implanted for long periods and permit long-term studies on unanesthetized animals.

The first important result of the present studies was the finding that the so-called thermoregulatory centers do not correspond to anatomically defined structures. Their constituent neurons are intricately connected with other neurons having other functions. These centers have been found mainly in the hypothalamus, but other parts of the CNS also seem to have the capacity to generate thermoregulatory responses.

6.6.2.1. Hypothalamic Centers

The first findings on the hypothalamic localization of thermoregulatory centers were made by Richet and Ott at the end of the nineteenth century. These findings have generally been confirmed by further investigations.

Generally speaking, lesions or destruction of the anterior hypothalamus inhibit the thermal response to a heat load while the response to cold persists. Conversely, injury to or destruction of the posterior hypothalamus inhibits the reaction to cold stress while leaving intact the response to warm.

Electrical (not thermal) stimulation of the anterior hypothalamus, mainly of the preoptic area, induces those coordinated responses that are normally observed during warm stress, i.e., behavioral reactions, skin vasodilation, and panting (in animals using this effector mechanism). Electrical stimulation of the caudal part of the diencephalon induces converse coordinated responses normally exhibited during regulation against cold, i.e., general vasoconstriction and shivering.

These experiments support the theory that there could be two thermoregulatory centers, anatomically separated but linked together by numerous connections. The first is located in the preoptic area of the anterior hypothalamus (see Figure 6.15). It is activated by a warm stress and triggers the heat loss mechanisms. It was therefore termed "thermolytic." Note that the same area of the hypothalamus contains central thermoreceptors as described above. The second center is also located in the hypothalamus but to the rear. It is activated by cold stress and was therefore considered the "thermogenic" center.

Today, however, these ideas seem too imprecise. Stimulation of a given part of the hypothalamus, for instance the preoptic area, does not always induce the same response, i.e., panting and peripheral vasodilation. On the other hand, selective destruction of a part of the hypothalamus does not totally suppress the capacity for response to the respective thermal stimuli, but does modify the threshold of the response to the opposite stress. For example, destruction of the preoptic area of the anterior hypothalamus partly inhibits the heat loss mechanisms and partly disturbs the mechanism of increased metabolism.

In man, hypothermia is most common when the hypothalamus is injured or destroyed, even if the pathological cause (e.g., angioma,

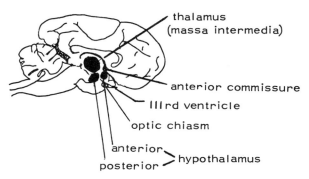

Figure 6.15. Sagittal section of dog brain, showing the thermoregulatory centers.

tumor) has affected the anterior part of the hypothalamus more than the posterior.

6.6.2.2. Other Thermoregulatory Centers

There are certainly other thermoregulatory structures in the CNS, mainly in the brain stem and the spinal cord. However, their roles are difficult to define, at least in normal life. In the highest species of mammals, the activities of these structures seem to be inhibited by the hypothalamic centers. However, in the lowest species of mammals, these secondary centers are apparently able to trigger coordinated thermoregulatory mechanisms, including behavioral responses.

There are few data on this subject for man. Some authors have described hypothermia following agenesis or destruction of the corpus callosum. However, surgical sectioning of this structure (performed in cases of severe epilepsy) does not produce a change in temperature regulation.

6.7. THE ROLE OF THE ENDOCRINE GLANDS

Apart from the adrenomedullary glands, the endocrines are not involved in thermoregulation as directly as the nervous system. They are not able to produce a thermal effector mechanism of their own, but can only modify the level of activity of other mechanisms. On the other hand, changes in the activity of a gland may sometimes be the result of a given thermoregulatory mechanism but not its cause.

The role of endocrine glands appears to be more important in cold than in heat regulation.

6.7.1. Adrenomedullary Glands

Through their secretory products adrenaline and noradrenaline, the adrenomedullary glands participate directly in cold regulation by enhancing and extending the effect of the sympathetic nervous system on the cardiovascular system. The products of these glands also effect thermogenesis in all tissues, but particularly in brown fatty tissue, where it exists. For this reason, the role of these glands will be described more fully in Chapter 7.

6.7.2. Thyroid Gland

For a long time it was accepted that the thyroid hormones, mainly thyroxin, played an important role in the thermal state of the body. Classically, thyroxin increases oxygen consumption by the tissues, and subsequently their heat production. Repeated injections of thyroxin tend to induce a rise in central temperature. The tendency to hyperthermia, on the other hand, is a classical sign of hyperthyroidism in the human. Conversely, hypothyroidism is accompanied by an increased sensitivity to cold and a tendency for the central temperature to decrease.

However, the actual role of the thyroid hormones in thermoregulation has to be reevaluated. For instance, cooling the hypothalamus (in the monkey) produces a thermogenic response. It had been postulated that this response was partly under the control of thyroid-stimulating hormone (TSH) and thyroxin. However, some investigations, e.g., those by Gale (1973) and co-workers, have shown that the release of TSH was not a constant result of the hypothalamic cold stimulus, and therefore that the increase in thyroid activity was not a prerequisite for obtaining a thermogenic response.

It has been suggested that thyroxin should enhance sensitivity to the thermogenic activity of catecholamines. It appears that this sensitizing action occurs even at the normal blood concentration of thyroxin, and that an increase in its concentration has no special complementary effect.

It had been generally accepted that the relative calorigenic effect of thyroxin was due to peripheral action, i.e., a direct action on the cells. However, another pattern of action has been observed. When injected centrally into the cerebral ventricles, thyroxin also induces a rise in central temperature. This central effect is probably very complex. As shown by Kaciuba-Uscilko et al. (1979), the hyperthermic effect of centrally injected thyroxin varies with the thermal state of the organism. It is low if the body is exposed to a warm ambience or has been injected with a pyrogen, but the effect is high during muscular exercise. In this case, centrally injected thyroxin could have a hyperthermic effect, especially by reducing heat loss.

6.7.3. Adrenocortical Glands

Exposure to a cold environment rapidly induces a rise in secretion by the adrenocortical glands. If cold exposure is prolonged, adrenocortical weight increases. The adrenal cortex is stimulated by many different forms of stress, e.g., severe pain or a high noise level. Therefore, its response to cold is probably nonspecific. The level of ascorbic acid in the tissues is lowered during the first hours of exposure to cold. If exposure is prolonged, the level thereafter gradually increases.

6.7.4. Pituitary Gland

The increase in activity of both thyroid and adrenocortical glands after exposure to cold is mediated by the hypophysis and its respective stimulating hormones, TSH and adrenocorticothyroid hormone (ACTH). The stimulation of the pituitary to produce these hormones is probably under the influence of the hypothalamus, acting via its releasing factors. The hypothalamus itself receives the

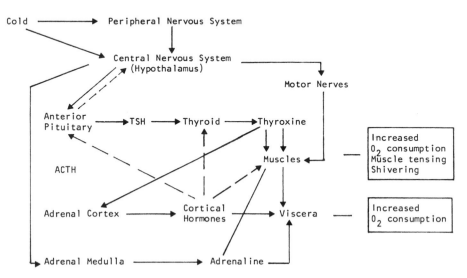

Figure 6.16. Hormonal response to cold exposure [from Hardy, 1967].

"cold" signals from the peripheral thermoreceptors and, eventually, the central thermodetectors.

Figure 6.16 is an attempt to summarize the endocrine reaction to cold exposure.

6.7.5. Posterior Pituitary

Exposure to heat is generally associated with a decrease in diuresis, whereas exposure to cold induces a diuretic effect. These modifications are, at least partly, under the influence of changes in the level of antidiuretic hormone (ADH).

These changes appear to be secondary to a modification in the distribution of blood volume, or in the total blood volume itself. During cold exposure, vasoconstriction occurring mainly at the extremities induces an increase in central blood volume. The intrathoracic "volume" receptors act in return by decreasing the release of ADH. During heat exposure the inverse phenomenon occurs. Moreover, if the water supply is not adequate, subsequent dehydration of the body is a complementary stimulus for releasing ADH. Therefore, changes in the level of ADH during thermal stress appear to be linked more to events produced by stresses than directly to the stress itself.

Some experiments, however, have indicated that the response of the posterior pituitary to heat stress could be very rapid, occurring before any change in blood volume distribution, and indicating a direct action of the thermal centers on ADH secretion or release.

6.8. MODELS OF THERMOREGULATION

Like other processes, thermoregulation can be represented by a diagram showing the morphological elements and their functional relationships. The term *model* can be applied to any such schematic representation. However, it is more appropriate to use this term only in reference to quantitative calculations, which may be either physical or mathematical.

One kind of model of thermoregulation is primarily descriptive. It illustrates how heat flows from the body to the environment and the measurement of such exchange. If founded on experimental data, such a model can be subsequently used to predict the response of the body to a given thermal stress. One of the best examples of this type of model is that proposed by Stolwijk (1971) for use in aerospace research.

Another kind of model attempts to "explain" how the body achieves thermoregulation. These explanatory models are generally based on a comparison with technological regulatory systems. Originally such models were empirical and not quantitative. Today, they owe accurate mathematical expression to the principles of control theory.

At the end of the last century, Liebermeister (1875) compared the human thermoregulatory system to a thermostatically controlled bath. He postulated that there must be a reference temperature for the body. When the actual temperature deviated from the reference temperature, the body triggered its effector mechanisms in order to reset the temperature at its initial level. On this assumption, fever was no more than a modification of the reference temperature. The principal advantage of this hypothesis—or, rather, comparison—is its simplicity. A thermostatic bath and its operation are universally understood. However, human thermoregulatory function is far too complicated to fit this theory.

Let us pose two questions. The first question is very simple: Is the temperature of the body constant? The answer is also very simple: It is not.

In clinical practice, accurate temperature measurement of a patient requires special and almost artificial conditions, e.g., metabolic rate at basal level, thermal neutrality. Conversely, the temperature of a subject working in a relatively warm environment is higher. It has long been demonstrated that core temperature bears an almost linear relationship to the heat load to which the subject is subjected. Finally, cyclic variations (e.g., diurnal, estral, seasonal) are also known to occur. The obvious conclusion is that the deep body temperature of a normal subject is not maintained at a fixed value, but can vary to a moderate degree.

An attempt was made to answer this point by assuming that the reference temperature was variable, under the influence of such factors as ambient temperature or level of muscular exercise. This assumption appeared very attractive; however, it does not provide a solution to the problem. We can certainly state that body temperature varies. But to say that any observed variation is due to a change in temperature set point does not provide additional information, because the set point cannot be measured in the first place. Furthermore, some authors who advocate this theory have a tendency to confuse the *actual* temperature of the body with the *reference* temperature. In a thermostatic bath, we are able to measure the actual temperature of the water and define the reference temperature; both temperatures are known. They are generally different, since it is this difference that triggers the regulating mechanisms. In the body, on the other hand, we may possibly be able to measure the actual temperature, but we are totally unable to measure or determine the reference temperature.

Our second question concerns the *level* of temperature regulation. As shown in Chapter 5, each organ, or part of an organ, has its own characteristic temperature. The temperature range is small, especially in the heart and brain. Since the brain, and especially the hypothalamus, is the location for thermoregulatory centers, it was also thought to be the site of temperature regulation. Indeed, in some species, the basal part of the brain including the hypothalamus has a special cooling system. However, we can easily conceive of the brain as being thermally regulated more closely than other organs without assuming that it is the brain's temperature, and only this temperature, that is the "regulated value."

In fact, there are many experiments that show that (1) thermoregulatory responses can be elicited by changing the temperature of other areas of the body, e.g., the spinal cord, with the hypothalamic temperature remaining constant, and (2) the temperature of the hypothalamus can be modified without triggering the effector mechanisms, if another part of the body is submitted to a compensatory change in temperature. Therefore, although the hypothalamus and in fact the entire brain are especially protected against large temperature changes, their temperature is not the only factor in regulation. Temperatures from all areas of the body are integrated by these centers and contribute to total thermal regulation.

It can thus be concluded that the comparison of the body with a thermoregulated system, such as a thermostatic bath, cannot be sustained. This comparison is too simple and is unable to take account of the complexity of the biological problem. The body is subjected to a variety of thermal stresses and its thermoregulation must follow variations in the thermal load. The problem must be set in terms of dynamic regulation, i.e., heat flow rates.

A better comparison would be to a cistern containing water. Although the inflow and outflow can independently vary, the regulating system must follow these variations as rapidly and accurately

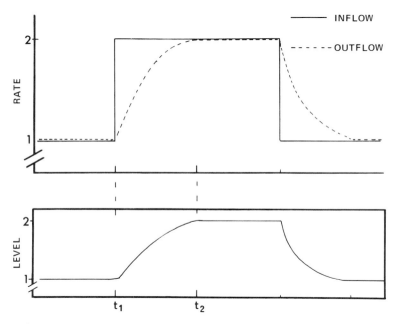

Figure 6.17. Dynamic regulation of the water level in a cistern. Initially the water level is constant at L_1, because inflow rate and outflow rate (I_1 and O_1 respectively) are equal. At time t_1, there is an abrupt disturbance, for instance a rise in the inflow rate to the value I_2. The controller triggers the regulating mechanism, which can be an increase in the outflow. However, this mechanism is not immediate. During this transient period, in which the inflow rate is higher than the outflow rate, the water level increases progressively. At time t_2, the output has become equal to the input rate and a new steady state is now obtained. The water level is constant again, but at a higher value than initially. It will return to its initial value when an inverse disturbance occurs (rightmost third of the figure). The regulation of body temperature may be expressed in a similar manner (see the text for further explanation).

as possible to maintain the level of water within a narrow range. Such a system of regulation is common in technological applications (Figure 6.17).

If we consider this cistern as in a steady state, the inflow rate I_1 is equal to the outflow rate O_1 and the level of water is constant at L_1. If the inflow rate abruptly increases to a value I_2, the equilibrium is broken and the level of water progressively rises. Let us now assume that regulation occurs, acting, for example, to increase the complementary total output to a new value O_2. As long as this result is not yet achieved, i.e., as long as the outflow rate is less than the value O_2, the level of water continues to increase. When the balance is achieved again, the level of water does not increase further. However, this final level L_2 is higher than the initial level. An opposite disturbance will have an opposite effect on the level, and, if the disturbances can statistically occur in both directions, the long-term result will be that the water level will vary over a limited range.

Some recent experimental work has considered bodily thermo-regulation from this point of view, and especially its dynamic aspects. A subject is placed in a thermoneutral environment and is in a steady state. He is then abruptly submitted to a decrease in his heat loss by a rise in the ambient temperature. The chronological study of events occurring after such a change provides interesting evidence of the mechanism by which the body's thermal controller integrates thermal signals and triggers regulatory processes. This study shows that the body responds to an unbalanced state by progressively utilizing its evaporative mechanisms. As long as the rate of evaporative heat loss does not compensate for the imbalance, there is a progressive increase in body temperature. At the end of the experiment, when the new steady state is achieved, the final temperature is now increased above its initial level.

A general scheme for thermoregulation would therefore be as follows (Figure 6.18). The temperature(s) of the body is/are obviously the result of the balance between heat production and heat loss, just as, in a cistern, the water level is the result of a balance between inflow and outflow. As long as the inflow and outflow of heat are equal, the level of temperature remains constant. If one of the two varies, there is then a progressive negative or positive heat storage and a subsequent change in body temperature. By their static

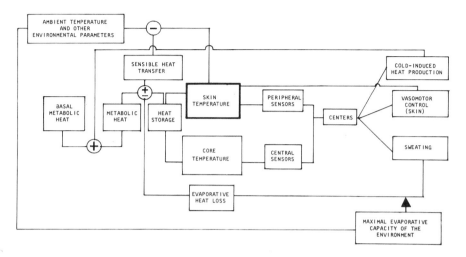

Figure 6.18. Schematic model of thermal balance as a complex servocontrolled system. Metabolic heat and sensible heat transfer constitute the main input and output of the system, respectively. In neutral conditions metabolic heat is at its basal level and compensates exactly for sensible heat transfer. Heat storage is then at zero. By the laws of thermodynamics this balance is observed at an ambient temperature of about 30°C, a skin temperature of 33.5°C, and a deep body (core) temperature of 37°C.

If there is an imbalance between input and output, heat storage changes from zero, which in turn results in a change in skin and/or core temperature. The respective thermosensors are activated, and their impulses are integrated by the thermoregulatory centers, which trigger the necessary effector mechanisms. These are the skin vasomotor systems, which modify skin temperature, and either cold-stimulated heat production, which adds to the basal metabolic rate, or sweating, an efficient means of heat loss. Evaporative heat loss, however, is dependent on the evaporative power of the environment, and also induces cooling of the skin.

Physiologically speaking, skin temperature, core temperature, cold-stimulated heat production, and evaporative heat loss may be considered as outputs of the system. Particular attention may be focused on one or more of these parameters, according to the investigator. For example, evaporation is important in studies of heat stress and heat tolerance; skin temperature, in studies of comfort; and core temperature, in studies of temperature regulation in mammals [from Mitchell *et al.*, 1972; Houdas *et al.*, 1973; Webb *et al.*, 1978; Werner, 1980].

and dynamic responses, the thermoreceptors located in many different sites within the body inform the thermal centers of the progressive temperature changes. The centers integrate these signals and trigger the appropriate mechanisms. These mechanisms (thermogenesis or evaporative heat loss) must exactly compensate for the imbalance in heat loss that has occurred. This compensation is not immediate. When it is achieved, a new thermal balance, and therefore a new steady state, is achieved. The progressive variation in body temperature is stopped, but, in absolute terms, the new temperature(s) is/are now different from the initial one(s). Therefore, the variation of body (and mainly central) temperature(s) is easily explained by the time constant of the regulatory mechanisms; the more rapid these mechanisms, the more limited is the temperature increase.

Although this explanation requires supporting evidence, it takes account of all known experimental data and gives a logical explanation of the observed facts without the need for a further hypothesis. This explanation correctly fits that given for fever in Chapter 9.

Heat Loss and Conservation

7.1. THE CARDIOVASCULAR REACTION

It has been shown in a previous chapter that the value of skin temperature at a given point is the result of an equilibrium between heat received from the body core, mainly via the blood, and heat lost into the environment. In turn, skin temperature is one of the most important factors determining the transfer of detectable heat. Skin circulation, which markedly affects skin temperature, is itself modified by a feedback system. Finally, variations in blood circulation to the skin produced by changes in skin temperature can affect cardiovascular function.

There is another condition in which the cardiovascular system is submitted to stress that can be very important: muscular exercise. However, muscular exercise is also a condition in which heat production can be dramatically increased. The combined effects of thermal stress added to work stress act together on the cardiovascular system. For this reason, the effects of thermal stress will be considered separately, first in the resting and then in the exercising subject.

7.1.1. General Considerations

Heat produced by the body core tissues must be lost into the environment. It has been shown in Chapter 4 that at least two stages can be described in this transfer: (1) transfer between body core and skin and (2) transfer between skin and ambience.

7.1.1.1. First Stage

Between body core and skin, heat is transferred in two ways: by conduction from tissue to tissue and by blood convection (blood displacement). Blood is a poor conductor of heat but can be included in a formula that describes the transfer of heat by circulation. The rate of metabolic heat transfer from the core at temperature T_d to the skin at temperature T_s, $\dot{H}_m$, may be expressed by the following equation:

$$\dot{H}_m = h_b(T_d - T_s)A$$

where A is the surface area of the body and h_b is the "coefficient of convection." This represents the mean heat flow through the skin surface per degree of gradient fall in skin temperature. It has often been referred to as "conduction"; however, as seen above, the major part of heat transfer is effected by blood convection, and the term *conduction* should be avoided.

In the simplest model, the body is considered as a core covered by a shell. This shell receives the arterial blood flux $\dot{Q}_s$ at temperature T_{art}. The venous blood flowing out to the skin has lost heat and the venous temperature, T_{ven}, is then lower that T_{art}. Therefore

$$\dot{H}_m = \dot{Q}_s \rho_b c_b (T_{art} - T_{ven})A$$

where ρ_b and c_b are the density and the specific heat of the blood, respectively. If it is assumed that T_{art} is close to T_d and T_{ven} to T_s, it appears now that h_b can be expressed by the product $\dot{Q}_s \rho_b c_b$.

7.1.1.2. Second Stage

The second stage demonstrates the transfer of heat from the skin to the environment. In the absence of evaporative heat transfer, this second stage can be described by the following formula:

$$\dot{H} = h(T_s - T_a)A$$

where $\dot{H}$ is the sensible heat flow and h is the combined coefficient for heat exchange (see Chapter 1). In a steady state, the heat exchanged between the skin and the ambience corresponds exactly to that arriving at the skin from the body core.

It is now possible to establish all the factors controlling skin temperature. They are summarized in Figure 7.1. Ambient temperature, T_a—and other factors, including the air velocity, V—determine the rate of heat lost by the skin and thus act directly on the skin temperature level. On the other hand, body core temperature, T_d, influences T_s, both directly (by tissue conduction) and especially indirectly (by blood convection). The amount of heat transferred by the blood depends on the amount of blood that reaches the skin. This is closely related to the level of vasomotor tone, and affects the convection coefficient h_b. Since T_a is set by the environmental

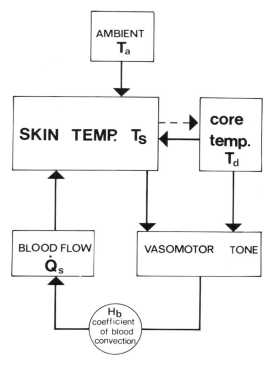

Figure 7.1. Scheme of factors contributing to skin temperature.

conditions and T_d is almost constant, the body can control its exchange of detectable heat by modifying skin blood flow and then changing the vasomotor tone of its vessels. For example, if cutaneous vasodilation occurs, $\dot{Q}_s$ is increased and the following phenomena are successively produced:

☐ Increase in heat flow to the skin
☐ Increase in skin temperature
☐ Increase in the skin-to-environment thermal gradient
☐ Increase in the heat lost by the skin and body

Vasoconstriction will produce the reverse reactions.

As shown in Figure 7.1, the control of skin vasomotor tone itself results from a loop mechanism triggered by signals issuing from the skin and from the body core, mainly the hypothalamus. Thus the level of skin vasomotor tone depends not only on the temperature of the skin but also on that of the core.

Although the entire skin surface participates in heat exchange, the limbs, and particularly the extremities, play the major role because of their structure (small volume with large surface area). However, the temperature of the blood arriving at the small vessels of the limbs is not equal to core temperature because of countercurrent heat exchange. This problem must be considered before the influence of T_s and T_d on vasomotor tone can be evaluated.

7.1.2. Countercurrent Heat Exchange

In the limbs, arteries and veins are often in close contact. This configuration causes the heat arriving via the arterial blood from the core to be transferred by conduction to venous blood, flowing from the extremities back to the core. For this reason, the temperature of arterial blood progressively decreases toward the extremities, resulting in a fall in skin temperature and a subsequent decrease in heat loss.

Total body heat exchange depends on the way in which blood returns to the heart. In a cold environment, blood return is provided by the deep veins, which are in close contact with the arteries. In this case countercurrent heat exchange is maximal, leading to conservation of heat in the body. In a warm environment, blood returns to the heart mainly by peripheral veins. In this case, countercurrent heat exchange is low and more heat is lost into the ambience.

In spite of this, the most of the regulatory mechanisms of the cardiovascular system normally occur in small vessels. These mechanisms are triggered by local changes in environmental temperature (which primarily alter skin temperature) and body core temperature.

7.1.3. Skin Blood Flow and Local Temperature

The influence of local temperature on skin blood flow is generally demonstrated in the hand for three reasons: (1) it is technically simple, (2) skin represents a significant part of the mass of the hand (50% on average), and (3) vascular reactions to temperature are more marked at the extremities than in the trunk. Typically the hand is put in a plethysmographic bath, and the water temperature is changed. Hand blood flow changes rapidly, within a few seconds, but then remains constant for a given temperature. Owing to the experimental procedure, skin temperature is generally close to the water temperature. Therefore a graph representing hand blood flow versus hand skin temperature (or water temperature) (Figure 7.2) can be obtained.

When the skin temperature of the hand is between 28° and 35°C (corresponding to the temperature exhibited by the skin in air of temperature 17°–27°C), blood flow changes little, varying from 1 ml $\cdot$ min^{-1} to 3–4 ml $\cdot$ min^{-1} per 100 cm^3 of tissue. At low temperatures, blood flow decreases slightly, up to about 0.5 ml $\cdot$ min^{-1}. However, if the cold stress is severe, for example in water at less than 15°C, hand blood flow tends to rise. This phenomenon is due to the so-called "paradoxical vasodilation" (see Section 7.1.3.2). The most marked variations in hand blood flow are observed in a warm environment. Above a skin temperature of about 35°C, blood flow increases, and at a skin temperature of 45°C, which is the maximum tolerable value, the flow can be about 30–35 ml $\cdot$ min^{-1} $\cdot$ 100 cm^{-3}. Nevertheless, it does vary from one subject to another, and over time in the same subject. For this reason, blood flow can, but only transiently, overshoot the mean value reported above, and reach peak values of up to 75 ml $\cdot$ min^{-1} $\cdot$ 100 cm^{-3}.

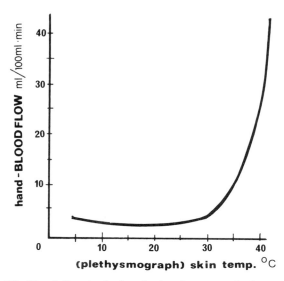

Figure 7.2. Blood flow in the hand related to water bath temperature.

If the entire skin surface is subjected to a warm environment, the general vasodilation that occurs can markedly affect cardiac activity. It must be noted, however, that in everyday life skin temperature rarely exceeds 36°–37°C.

The vascular reactions of the hand appear to be homogeneous. However, these reactions are chiefly confined to the fingers. For 100 cm^3 of finger tissue, blood flow at thermal neutrality is 10 ml · min^{-1} · 100 cm^{-3}. It decreases to 1–2 ml in a cold ambience but can reach 100 in a warm environment, with transient peaks of 150–180 ml · min^{-1} · 100 cm^{-3}. Conversely, the blood flow in the forearm exhibits a smaller range of variations, from 2–15 ml · min^{-1} · 100 cm^{-3}.

The thermal reactivity of the vascular bed in a lower limb is less than that in the upper limb; however, within the limbs a similar difference exists, extremities being more reactive than thighs. On the face, this reaction is more moderate, and it is minimal on the remaining skin surfaces.

7.1.3.1. Secondary Adjustments

The vasodilation of skin vessels when the body is heated also produces a shift in blood volume. The rise in skin blood flow is then associated with a marked increase in cutaneous blood volume.

There is, however, a difference according to whether the subject is supine or in an upright position. In the supine position, the hydrostatic forces on the distribution of blood are minimal and the shift in blood volume is moderate. In the upright position, more than 55% of the blood volume is in the veins below the heart. Therefore, when the body is heated, there is not only a vasodilation in the small cutaneous arteries, but also a rise in the distensibility (compliance) of the limb veins, leading to an important increase in the blood volume of the limbs (Figure 7.3). This explains how discomfort or even pain can occur when subjects are in an upright position in a warm room, especially if the heating is through the floor.

Other modifications during heating of the body are also observed. While skin venous volume increases, return of blood to the heart decreases proportionally. In the first moments, the right ventricle can maintain its output by depletion of its sump, that is, the thoracic and splanchnic veins. There is thus a fall in the central venous volume. On the other hand, an increase in cardiac output is

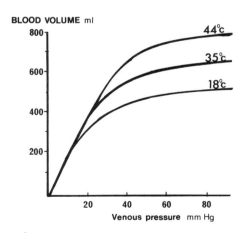

Figure 7.3. Increase in blood volume with increasing venous pressure at different temperatures.

possible only if the left ventricle withdraws blood volume from the
pulmonary vessels. However, arterial pressure cannot be maintained
despite visceral vasoconstriction, and it decreases slightly.

As we have seen previously, the blood flow of, for example, the
hand, closely depends on local skin temperature. It also depends on
the mean skin temperature, and consequently on the temperature of
the surrounding ambience. The higher the ambient temperature, the
more marked is the reaction of hand blood flow to local temperature.
However, this reaction is also influenced by the central temperature:
For a given skin temperature, the higher the internal or core temperature, the higher will be the skin blood flow.

It appears then that cutaneous vasomotor tone is under the
control of signals issuing from thermoreceptors located both in the
skin and in the core. The study of the respective roles of these signals
constitutes an important part of an examination of total thermoregulation. On the upper extremities, which are generally exposed, skin
blood flow appears to be primarily under the influence of local
temperature, while internal signals play the major role in the control
of the skin blood flow of the other regions.

The thermal conditions of the scrotum present a particular
problem. In most species, and particularly in man, the testicles are
outside the abdomen and contained in the scrotum. The spermatic
artery is in very close contact with the corresponding veins, suggesting the existence of an important countercurrent heat exchange. The
scrotum has a large skin surface area, which can be strongly modified
by nerve reflexes. In neutral conditions, the temperature of the
testicles does not exceed 34.5°C, and that of the scrotal surface,
32°–33°C. Temperature regulation of the testicles is probably quite
precise. It is well known that even moderate hyperthermia of this
region results in an abnormal development of the spermatozoa.

Infrared thermography can be an effective aid to the diagnosis
of such vascular abnormalities as varicocele.

7.1.3.2. Paradoxical Cold Vasodilation

When skin temperature falls below 15°C, vasodilation occurs
and the blood flow increases locally, resulting then in an increased
heat loss. The term *paradoxical* is used to characterize this reaction.

If cold exposure is prolonged, periods of vasoconstriction and vasodilation alternate; this phenomenon is called *hunting*. Paradoxical
vasodilation is probably also exhibited by the muscular blood vessels.

The purpose of this vascular response is apparently to protect
the skin and subcutaneous tissues against marked cooling, which
could result in frostbite. If a cold stimulus sufficient to produce
paradoxical vasodilation is applied to the hand, and if at the same
time blood flow is suppressed in the limb by a cuff, pain rapidly
appears. It disappears when the cuff is removed and blood flow is
restored, thus rewarming the hand.

7.1.3.3. Factors Involved in Skin Vascular Response to Heat

The increase in the skin blood flow in response to heat depends
on certain key factors:

  □ A peripheral factor—vasodilation of the skin vessels
  □ A change in the distribution of the cardiac output
  □ An increase in the cardiac output itself

Mechanisms of Cutaneous Vasodilation. There are no vessels in
the epidermis, all cutaneous vessels being located in the dermis.
These vessels originate in the vascular hypodermal network, which is
at a depth of about 2 mm. In the dermis the small arteries form a
subepidermal plexus, which is at a depth of about 1 mm. The
terminal arterioles are directly followed by the metarterioles, from
which the capillaries directly originate. Veins form two anastomotic
networks in the dermis and a deep venous plexus is located in the
subdermal region.

The main characteristic of the skin circulation is the presence of
numerous anastomotic structures between small arteries and small
veins. Each anastomotic vessel is loop-shaped and it is called a
glomus. The glomera are innervated by the sympathetic system,
which controls their diameter. The diameter can vary from 0 to about
70 μm, allowing blood either to pass through the capillary bed or to
short-circuit it. When the anastomoses or shunts are open there is an
increased blood flow through the skin. The number of glomera varies

Table 7.1. Number of Arteriovenous Anastomoses (Glomera) per cm^2 of Skin at Different Locations within the Arm

Location	Glomera/cm^2
Hand	
Index finger	
Under nail	500
Fingertip	235
Third phalange (palmar)	150
First phalange (palmar)	95
Palmar surface	
Thenar eminence	115
Hypothenar eminence	95
Dorsal surface	Nil
Forearm	Nil

from one region to another but they are mainly located at the extremities (Table 7.1).

The existence of these anastomotic vessels has led to the assumption that they were the original site of warm vasodilation. Vasodilation can probably occur at the level of the capillary bed itself by the opening of the precapillary sphincter, because it is observed at a low level in regions where glomera are absent. However, vasodilation occurs mainly in the regions where the glomera are present: the greater the number of these structures the greater is their ability to vasodilate.

The mechanism of vasodilation is mainly nervous. It corresponds to the decrease, or even suppression, of the vasoconstrictor tone that the sympathetic system always imposes on the smooth muscle of the vessels, and especially of the glomera. The inhibition of this tone, by section of the sympathetic fibers or pharmacologically by α blockage, produces a significant vasodilation, similar to that observed when heat is applied. This neural mechanism explains the general vasodilation that occurs when the whole body is subjected to warmth or during muscular exercise. However, where only a small cutaneous region is heated vasodilation will appear only at this site. This local response has generally been explained by an axon reflex.

Humoral mechanisms could probably play a role in vasodilation. The vasodilation obtained by a total inhibition of the sympathetic system is not maximal, and it can overshoot if the thermal stimulus is intense. Some investigators, mainly Fox and Hilton (1958) and Randall (1953), have concluded that heat vasodilation is partly mediated by substances that appear in the blood during the activation of sweat glands. When activated they produce sweat on the external side and a proteolytic enzyme on the internal (circulation) side. This enzyme, called *kallikrein*, acts on a decapeptide normally present in blood plasma, *kininogen*, resulting in the formation of an octapeptide called *bradykinin*. Bradykinin has an important vasodilatory effect.

In a cold environment the response of vascular smooth muscle is apparently simple. Cold induces vasoconstriction owing to an increase in sympathetic tone. This vasoconstriction leads to a fall in skin blood flow and subsequently in skin temperature. α-Sympathetic blockade suppresses this vasoconstrictor reaction.

7.1.4. Cardiac Output

Figure 7.4A shows how cardiac output is distributed in local circulation. At thermal neutrality the skin circulation receives less than 10% of the total cardiac output. In warm conditions the fall in skin vascular resistance causes blood to pass more easily throughout the cutaneous vessels. Under these conditions skin blood flow can reach very high values, up to 6 liters $\cdot$ min^{-1} according to Rowell (1974). Such a value, however, implies not only a modification of the blood distribution, but also an increase in total cardiac output. Figure 7.4B shows how a cardiac output of 10 liters $\cdot$ min^{-1} is distributed during exposure to an intense heat stimulus.

Although the left ventricle increases its output, inverse modifications in local distribution occur, mainly in the splanchnic circulation, where blood flow may decrease by up to 35%. This is due to a local vasoconstriction, which is a part of the total response against a heat stress. Renal blood flow also decreases, but to a lesser degree, i.e., by about 25%. However, the cerebral, myocardial, and muscular circulations are not affected and remain normal.

Changes in Cardiac Output. It has been shown that cardiac output itself is raised if the thermal stimulus is strong enough.

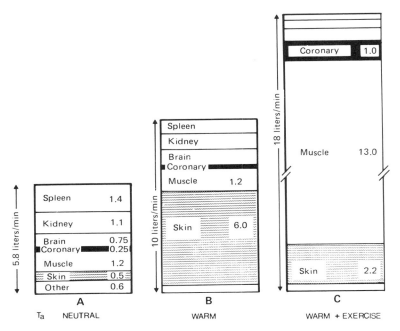

Figure 7.4. The regional distribution of cardiac output under neutral, warm, and warm + exercise conditions.

Cardiac output may increase by up to 75% of the basal level. This increase is due primarily to the rise in frequency, since stroke volume does not change markedly (increasing maximally by only 5%).

The increase in cardiac output does not compensate exactly for the decrease in vascular resistance, so that arterial blood pressure tends to fall slightly. However, increased cardiac activity implies increased myocardial work, which explains the low tolerance to warmth of patients with cardiac insufficiency. In cold conditions the first response of the cardiovascular system is the vasoconstriction of skin vessels. The second and more adaptive response is an increase in thermal metabolism. These two responses act on the heart in opposite ways:

□ A decrease of the peripheral blood flow tends to decrease cardiac activity.

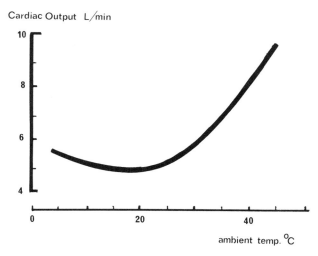

Figure 7.5. Cardiac output in relation to ambient temperature.

□ A rise in the oxygen consumption of the body because of its increased metabolism implies a corresponding rise in çardiac activity.

The competition between these two responses explains why the cardiac output starts to decrease (decreasing maximally by 30%) down to an ambient temperature of about 15°C but goes up at lower temperatures, so that the neutral level of cardiac output is reached again at about 5°C (Figure 7.5).

All these cardiac modifications are under the control of the sympathetic system acting on the β-receptors of the heart. They constitute a part of the whole response of the thermoregulatory system.

When the subject is heated externally, the rise in heart rate begins very rapidly after the onset of heating, and, as pointed out by Kerslake and Cooper (1950), this can be at a time when the hand blood flow is still at its resting level. This increase in heart rate appears to be due to a neural link between the peripheral thermal receptors and the nerves controlling the heart via the hypothalamus. However, the closest relationship is found between the body core temperature and heart rate. This link is mainly neural, originating in the central thermal receptors. There is, however, a direct action of the blood temperature on the cardiac side.

7.1.5. Muscular Exercise

Muscular activity has two major consequences for the cardiovascular system:

1. A significant increase in muscular metabolism requires a corresponding rise in blood supply to the muscles and therefore
   a. A change in the distribution of the cardiac output toward the muscular vessels.
   b. An increase in cardiac output itself, which can be increased four- to fivefold.
2. Increased muscular metabolism produces a significant amount of heat, resulting in an increase of heat production over basal levels of up to tenfold. Muscular activity, therefore, represents a major thermal stress.

Under cold conditions the two stresses act in opposition. Skin vasoconstriction allows more blood to flow toward the muscular vessels. The increase in thermal production also compensates for the increase in heat loss from the body. Under these conditions the stress on the heart is moderate. However, if the muscular activity is intense, increased heat production can be higher than the increase in heat loss. Even in a cold environment strong muscular activity can lead the body to trigger regulatory mechanisms against heat load. Therefore, the concept of a "cold environment" is related to the level of muscular activity the subject performs.

Under warm conditions the stress of muscular activity and that of warmth are additive in their effects. Significant heat production, which must be lost into the environment, tends to divert blood toward the periphery. At the same time, the active muscles require a considerable amount of blood. The heart can increase its activity only up to a maximal value. When the two stresses are intense, blood supply to the skin and to the muscles is achieved only at the expense of other regional circulations, e.g., that of the spleen (Figure 7.4C). However, there are limits to the possible increase of cardiac output and the bypass of blood to skin and muscles. Rowell (1974) found that, during maximal exercise, cardiac output did not change markedly with the subject at thermal neutrality or in a warm ambiance. Under these conditions, however, the contradictory needs of the skin and muscle result in competition between them. Muscular exercise is an indirect stimulus for cutaneous vasoconstriction and thus tends partially to counterbalance the skin vasodilation owing to warmth. On the other hand, muscular contraction tends to empty cutaneous veins. For these two reasons venous volume at a given core temperature is lower during exercise than during rest.

The main thermoregulatory mechanism under these conditions is the secretion of sweat followed by its evaporation. Sweat production corresponds to a loss of water and solutes and thus modifies blood volume, adding a further stress to the heart.

Finally, thermal balance requires that sweat must be evaporated, and this phenomenon depends on the air moisture. If the ambient humidity is too high, sweat evaporation may be impaired. Thermal equilibrium of the body cannot, therefore, be achieved.

These facts explain why work in hot, humid conditions represents maximal stress on the cardiovascular and thermoregulatory systems.

7.2. HEAT LOSS BY SWEATING

When the heat gain is higher than the heat loss, the first, or at least the more rapid, response of the body is a rise in skin temperature (see Chapter 6). This rise allows an increased loss of sensible heat. If this response is not enough to achieve thermal balance, an increase in heat loss can be effected only by raising the insensible heat loss, that is, the evaporative heat loss.

In humans this is achieved by the secretion and evaporation of sweat. As seen in Chapter 6, the heat that must be lost to achieve a new thermal balance has to be equal, physically speaking, to the value of the heat storage. Therefore, we must have

$$\dot{E} = \dot{S}$$

and

$$\dot{H}_e = \dot{H}_m + \dot{E}_b + \dot{E}$$

The latent heat of vaporization of sweat is close to that of water itself: 2.45 kJ $\cdot$ g^{-1} of water (0.585 kcal). Heat is removed from the medium in which the vaporization occurs, and in this way the medium tends to be cooled. Therefore, the vaporization of sweat must occur within the peripheral layers of the skin, or on the skin itself, in order to cool the body. If evaporation cannot be completed within these structures, sweat flows over the skin, resulting in a corresponding absence of evaporative heat loss.

Cooling of the body by sweating implies two mechanisms:

1. The production and secretion of sweat. This mechanism is essentially biological.
2. The evaporation of the sweat produced. This is mainly dependent on the physical characteristics of the ambient air and is thus mainly physical.

It is interesting to note that the physiological function by which the body may lose heat by evaporation varies according to species. The primates, the equidae, the bovidae, and others have a sweating system similar to that of humans. The dog and numerous other species lose heat by increasing their ventilatory volume (panting or polypnea), which produces a rise in the evaporative heat loss via the respiratory tract. Thermal polypnea has been described in the newborn and young infants. However, except in the premature, the rapid development of sudoral activity soon allows the infant to achieve its thermal balance by sweating. On the other hand, hyperventilation is frequent in infants and may be a cause of increased heat loss. Yet another mechanism is used by other species like the rat, which achieves body cooling through the evaporation of saliva deposited on the skin when it licks itself.

When a subject is subjected to an emotional stress, a burst of sweat appears promptly. This "psychologically" induced sweating is confined mainly to particular skin regions: the palms, the soles, and the axillae. Psychologists measure the electrical impedance of the skin, which is transiently decreased when such a sweat burst occurs, in order to quantify psychogenic stress. The duration of this secretion is short and the amount of sweat produced is negligible. Therefore, this form of sweating does not reach thermoregulatory significance.

7.2.1. The Sweat Glands

Sweat is produced by the *sudoral glands*. These glands are classically divided into two groups:

1. The *apocrine glands*, which are located only in certain areas, e.g., the axilla, the areola, the mons pubis, and the perianal area. The secretory function of these glands begins at puberty and becomes virtually continuous, although slightly modified by thermoregulation. The mechanism of secretion of these glands is poorly understood. Owing to their location, little of the sweat produced by these glands is lost by evaporation and most remains on the skin surface. The rate of apocrine sweat production varies from one subject to another but it is generally low: about $1 \text{ g} \cdot \text{m}^{-2} \cdot \text{h}^{-1}$ (Brebner and Kerslake, 1960).

2. The *eccrine glands*, which are the true regulatory glands. They are distributed over the entire skin surface, although unevenly. The forehead, the external face of the thigh, the palm, and the sole are generally rich in these glands, but there are again important differences between individuals. The total number of eccrine glands is about $3-4 \times 10^6$ and their total weight is 100 g (the same as the weight of one kidney). They first appear in the $3\frac{1}{2}$-month-old fetus, on the volar surface of the hands and feet, and later appear in other areas of the body. The eccrine glands consist of at least two parts (Figure 7.6A, B*):

   a. *Secretory tissue*. This is a coiled tube located in the dermis and composed of three distinctive cell types: clear, dark, and myoepithelial cells. The clear cells contain abundant mitochondria and form intercellular canaliculi with other neighboring clear cells. It is generally believed that the clear cells are responsible for the precursor sweat secretion

*Figure 7.6 is included in the color insert that follows p. 42.

produced from blood plasma. The dark cells produce mucus. The myoepithelial cells of the coiled portion probably have a supportive but not a contractile function.

b. *Ducts*. These consist of both a coiled portion and a straight ascending part. The presence of large myoepithelial cells at this level indicates that this part probably has a contractile function. However, contraction is not a prerequisite for inducing and maintaining sweat secretion (Sato, 1974).

c. *Transitional zone*. There is disagreement about the existence of a transitional zone between the secretory and the ductal parts. This zone was first described in 1949. After some contradictory results, it appears from the investigations of Ochi and Sano (1972) that this transitional part has no specific characteristics, and is structurally a mixture of the two major parts. These authors account for the discrepancies among previous results by the confusion between the so-called "transitional portion" and structures that "are nothing other than small vessels in the vicinity of the sweat glands." These vessels differ from the vessels seen in other organs, hence the confusion.

7.2.2. Sweat Production

Earlier work by Bulmer and Forwell (cited in Sato, 1974) suggested that sweat was produced by ultrafiltration of the blood plasma, resulting in a precursor fluid. The removal of some Na^+ ions and other substances by the duct cells resulted in the final secretory fluid. This hypothesis has been confirmed by more recent studies. The mechanism by which secretory cells of the coiled portion produce the precursor fluid is probably ultrafiltration. However, active mechanisms must be involved, at least in part because glucose appears to be necessary for the production of precursor fluid. The removal of glucose from the incubation medium of *in vitro* gland preparations abolishes sweat secretion. Although glands may store glycogen, the amount is too small to sustain a continuous rate of sweat secretion when glucose is removed.

Precursor sweat contains the same substances at the same concentrations as are found in plasma. However, when it appears at the skin surface it has a different composition and is always hypotonic compared with plasma. This is the result of a selective reabsorption of sodium (Na^+) by the ductal cells, without a corresponding reabsorption of water. Na^+ reabsorption is achieved by an active mechanism, against an electric potential difference of -100 to -120 mV. This Na^+ reabsorption explains a classical phenomenon. When the rate of sweat secretion is low, its Na^+ content is also low, and sweat is very hypotonic. This is because sweat flows slowly along the ducts, permitting increased reabsorption of Na^+. Conversely, if the sweat secretion rate is accelerated, the reabsorption is less effective, resulting in a relatively high concentration of Na^+ in the final secretion, which is slightly hypotonic.

The reabsorption of sodium by the duct cells is controlled by the adrenal mineral corticoids, which enhance reabsorption, resulting in a fall in the Na^+ content. However, the mechanism by which mineral corticoids accelerate Na^+ reabsorption is probably different from that observed in the tubular part of the nephron. On the other hand, mineral corticoids, and particularly aldosterone, reduce the rate of sweat production. It has been reported that local administration of antidiuretic hormone (ADH) reduces sweat production and increases Na^+ concentration. However, these responses are probably only due to the local vasoconstriction induced by ADH injection. When administered to the whole body, ADH does not appear to influence directly the water and electrolyte secretion of the sweat glands.

7.2.3. Sweat Gland Innervation

It is well established that the nerve structure of sweat glands is composed of nonmyelinated class C fibers of the postganglionic sympathetic fibers issuing from the paravertebral ganglia. The fibers are distributed to the myoepithelial and secretory cells. However, although these fibers are undoubtedly sympathetic, the chemical mediator is acetylcholine. Sweating is blocked by atropine, but it is evoked by acetylcholine and parasympatheticlike drugs. The presence of adrenergic fibers in the vicinity of sweat glands has been described in older observations, but these have been challenged. Yet recent observations on catecholamine-containing nerves around the sweat glands have again raised the old dual innervation theory. On the other hand, it has been observed that local administration of adrenaline was able to stimulate sweat secretion. It has been stated that cholinergic innervation is responsible for thermal sweat secretion, but adrenergic innervation affects "psychogenic secretion."

However, even this form of sweat production is probably also cholinergic. Experiments have been made by Ochi and Sano (1972) using the light microscopy formaldehyde-induced fluorescence method and a potassium permanganate method for the electromicroscopic detection of granular vesicles in the sympathetic nerves. These experiments showed that there are no granular vesicles that could be considered noradrenaline-storing structures. Although injected adrenaline can produce mild sweat secretion, it appears from these recent experiments that this action is humoral but not produced by adrenergic fibers. We may therefore conclude that the mechanism by which thermal and psychogenic sweating is produced is *cholinergic*.

7.2.4. Secretion Volume

Sweating is the mechanism by which the organism is able to lose heat that cannot be lost by convection, conduction, or radiation. Sweat secretion, therefore, depends on the thermal requirements of the body, providing that the necessary water is available. Sweat secretion is almost zero (except for the small production of the apocrine glands) in cold and neutral conditions. Secretion begins and grows progressively greater as soon as the heat load increases.

The determination of the maximum perspiration rate of the body presents a major problem. It is difficult to determine the duration of sweat secretion, which is an important parameter. A value of 2–3 liters $\cdot$ h^{-1} has been reported by Robinson (1963). These figures are very high and probably correspond to sweating of only short duration. If conditions imply that sweating must be maintained for several hours or more, a value of 1 liter $\cdot$ h^{-1} appears to be the maximum. Moreover, such a figure can only be maintained if corresponding amounts of water and electrolytes are supplied to the subject. When dehydration of the organism reaches 9–10% of body weight, sweating cannot be maintained at the required level; therefore, the thermal balance is destroyed and hyperthermia begins.

In a newborn baby, sweat glands do not develop. Maximal secretion is only one-third or one-half that of an adult. Maximal sweat production increases rapidly with increasing postnatal age.

7.2.5. Physical Properties and Chemical Composition of Sweat

7.2.5.1. Osmolarity

Sweat is always hypotonic compared to plasma. However, as already shown, this osmolarity is variable: The greater the sweat secretion, the higher the osmolarity, and vice versa.

7.2.5.2. pH

The pH of sweat is generally lower than that of blood. It may be as low as 5.0, but it increases with production to the neutral value. When exposed to air, sweat tends to become alkaline (pH 7–8).

7.2.5.3. Constitution

The major component of sweat, water excepted, is *sodium chloride* and it is therefore the major element determining osmolarity. Its concentration is very variable. Na^+ concentration can be as low as 5 mEq/liter at low secretion rates, and as high as 140 mEq/liter during increased secretion. This last figure is very close to that of plasma. Chloride (Cl^-) follows the variation of Na^+, the ratio Na^+/Cl^- being about 1 : 1 and almost constant. For a given secretion rate, acclimatization to heat induces a decrease in the NaCl concentration. This phenomenon is controlled by the adrenal mineral corticoids.

In contrast to that of Na^+, the concentration of *potassium* (K^+) decreases as the sweat rate increases, but always remains higher than the plasma concentration (5 mEq/liter). *Calcium* is present at a level that varies from 10 to 80 mg/liter. In its ionic form, its presence is essential for the secretion of sweat, as in many other secretory cells. "The influx into the cell of extracellular Ca^{2+} is an essential step in the events intervening between stimulation of the gland, and active secretion of ions and water by the secretory cells. The primary, if not the sole, role of acetylcholine may be the introduction of Ca^{2+} into the cell" (Sato, 1977). Other electrolytes are found in sweat, including magnesium (0.4–4 mg/liter), phosphorus (0.03–0.4 mg/liter), copper, and iron.

Organic substances are also found, but *glucose* is almost absent, occurring in concentrations of only 1–9 mg/liter. The mechanism by which glucose is reabsorbed is not yet totally explained. It probably differs from that observed in the nephron tubule because phloridzine does not affect the glucose content of sweat.

Lactate and *urea* constitute the major organic compounds secreted in sweat. Their concentration partly depends on the secretion rate. At low levels, lactate concentration is high (30–40 mmoles/liter), but it rapidly drops to a plateau at around 10–15 mmoles/liter as secretion increases. Acclimatization to heat lowers sweat lactate. The presence of lactate is probably due in part to the glycolytic activity of the secretory cells. In contrast, urea is not produced by the secretory cells but is derived from plasma urea. Sweat urea concentration is usually expressed as a sweat/plasma urea ratio. At a low secretion rate, this ratio is high (2–4), but it approaches a plateau at 1.2–2.5 as sweating decreases. The fact that this ratio is higher than 1 is thought to be due to a small reabsorption of water by the ductal cells.

Ammonia is also present, but at a concentration (0.5–0.8 mmoles/liter) that is 20–50 times higher than that in plasma. This concentration is inversely related to the rate of secretion and pH. The major part of this ammonia is derived from plasma by ionic trapping of NH_4^+ in an acid medium.

Amino acids are found in sweat at concentrations higher than those in plasma. Their concentrations appear to vary with the degree of exercise training. As pointed out by Chappuis *et al.* (1976), untrained subjects show significantly higher concentrations of alanine, arginine, glycine, histidine, and other amino acids, but not of citrulline, methionine, cystine and ethanolamine. However, the mechanism by which these amino acids appear in sweat remains unknown.

Human eccrine sweat also contains *proteins* at a low concentration, 20–75 mg/liter. The major proteins are albumin and alpha and gamma globulins. Fatigue may enhance the concentration of these proteins. No toxic molecules have been found in sweat, unless already present in the plasma. In classical experiments, death was observed after injection of sweat into small animals, but it was probably due to a microbial contamination or immunologic shock.

7.2.6. Patterns of Sweat Production

For any given gland, sweat production appears to be intermittent. It is possible that sweat production could be continuous but for the rhythmic contractions of the myoepithelial cells of the duct. An increase in the total volume of sweat produced is brought about both by increased glandular production and by an increase in the number of secretive glands secreting at a given time. At the onset of sweating, the first area to sweat is generally the forehead, followed in order by the upper arms, hands, thighs, feet, and, finally, back and abdomen (Figure 7.7).

The best experimental procedure to study the time sequence of sudoral activity is to submit a subject to a sudden increase in the

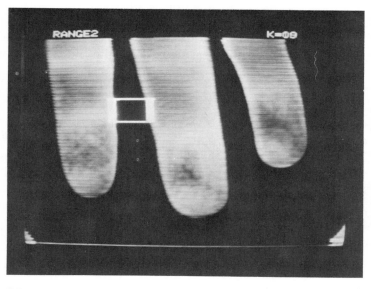

Figure 7.7. Infrared thermogram showing cooling at the fingertips by evaporation of sweat (black = cold).

environmental temperature. Under neutral conditions, sweating is practically absent. After an increase in ambient temperature, the onset of sweating is delayed for up to 20 min. After this delay, sweat secretion begins and increases progressively. This increase approximates an exponential curve. Finally, sweat secretion reaches the level corresponding to a thermal balance and stabilizes. Because of the exponential increase, it is possible to define a time constant for the sudoral response to increased heat load. This time constant is variable between subjects, its mean value being between 5 and 10 min. The existence of delay at the onset, and, later, of a time constant for sudoral activity, is very important in practice. When the thermal equilibrium is disturbed, a new thermal balance cannot be achieved for 45 to 60 min. This figure is still longer in newborn babies and premature infants. For this reason, thermoregulation against warmth is a very long process, and in practice a subject never reaches a true thermal steady state.

7.2.7. Evaporation of Sweat

The process of sweat production is only the first part of the complex mechanism leading to body heat loss. After its production, sweat must be evaporated and this process must take place at the skin surface. In this way, the heat needed for the transformation of liquid (sweat) into water vapor is taken from the skin and then from the body, which is consequently cooled.

To achieve thermal balance, the body must lose heat at rate $\dot{E}$, equal to the heat storage rate, $\dot{S}$. This rate of energy loss, $\dot{E}$, requires the production of sweat at rate $\dot{m}$. Using the subscript $_{req}$ (required) we define the necessary rate of heat loss ($\dot{E}_{req}$) and the corresponding sweat production rate ($\dot{m}_{req}$).

If the body can produce sweat at the required secretion rate, the main problem of the second part of the mechanism is to determine whether or not this sweat can be evaporated at the skin surface into the atmosphere. This evaporation depends mainly on the characteristics of the atmosphere. Chapter 4 showed that the evaporation rate from a water surface is related to the difference between the water vapor pressure at this surface and the water vapor pressure of the surrounding atmosphere. The first value depends only on the temperature of that surface, because it corresponds to the saturated water vapor pressure at that temperature, P_{ws}^0. Therefore

$$\dot{E} = h_e \left(P_{ws}^0 - P_{wa} \right) A_e$$

where h_e is the coefficient of evaporation, which depends on the convection coefficient h_c according to Lewis's relationship ($h_e = 2.25$ h_c), and A_e is the surface area of the water.

Therefore, if the skin were a totally wet surface, like water, the maximum evaporative heat loss rate would be $\dot{E}$. The subscript $_{max}$ can then be added to $\dot{E}$ to indicate that this value is the maximal theoretical rate for evaporative heat loss from the skin. It corresponds to a rate of sweat secretion $\dot{m}_{max}$ with $\dot{m}_{max} = \dot{E}_{max}/2.45$.

This concept is called the *maximal evaporative power (capacity)* *of the ambience.* It has two aspects, one material ($\dot{m}_{max}$) and another energy ($\dot{E}_{max}$).

$$\dot{E}_{max} = h_e \left(P_{ws}^0 - P_{wa} \right) A_e$$

$$\dot{m}_{max} = \dot{E}_{max}/2.45$$

The problem can now be described in the following terms. Assume an atmosphere in which the water vapor pressure is P_{wa} torr in absolute value. In this atmosphere, the body has a skin temperature T_s. According to the other climatic and physiological conditions, the subject must evaporate heat at rate $\dot{E}_{req}$. Therefore, the first requirement is to produce an equivalent sweat rate, $\dot{m}_{req}$. The second is to evaporate this sweat. The characteristics of the atmosphere determine the maximal possibility for evaporation, that is, $\dot{E}_{max}$.

If $\dot{E}_{req} \ll \dot{E}_{max}$, for instance if the required heat loss is 10% of $\dot{E}_{max}$, the sweat produced can easily be evaporated and thermal balance can be achieved. Note that in such cases the sweat being totally evaporated does not appear as drops on the skin and sweating is not apparent to the observer.

Conversely, where $\dot{E}_{req} \ll \dot{E}_{max}$, that is, where $\dot{m}_{req} > \dot{m}_{max}$, only one part of the sweat produced, corresponding to $\dot{m}_{max}$, can be effectively evaporated. The difference between $\dot{m}_{req}$ and $\dot{m}_{max}$ cannot be evaporated and appears then as liquid sweat flowing on the skin. Therefore, in this case heat is not removed from the body and thermal balance cannot be achieved.

Therefore, at least theoretically, thermal equilibrium should be achieved as long as $\dot{E}_{req}$ is less than $\dot{E}_{max}$, and liquid sweat might not appear until this limit. However, this is not the case in practice. On the one hand, liquid sweat appears on some skin areas although thermal balance is completely achieved. On the other hand, the limit of thermal homeostasis is reached far before $\dot{E}_{req}$ becomes equal to $\dot{E}_{max}$. It is accepted (e.g., Kerslake, 1972; Vogt, 1968) that the limit of thermal homeostasis is reached when $\dot{E}_{req}$ is about 40% of $\dot{E}_{max}$.

The discrepancy between theory and fact is due to the fact that the above reasoning was based on the body skin being a flat surface and the atmosphere homogeneous. This is not the case. Skin surface has a very complex shape, and the atmosphere is not the same for all skin regions. Moreover, local sweat secretion varies markedly according to skin area. The result is that (for example, on the internal face of the thigh) local air may have a lower maximal evaporating power than that of the atmosphere in general. Therefore, at this site, sweat produced cannot be totally evaporated. In other sites, such as the forehead, the production of sweat can be very high. Even if the evaporating power of the local ambience is important, all sweat cannot be evaporated. The variations in local ratios of sweat produced to evaporating power explain why a subject can exhibit sweat drops on some skin areas although a total thermal balance has been achieved.

Nevertheless, the subject loses more sweat than is necessary. Some authors have introduced the concept of *sweating efficiency* for describing this phenomenon as follows:

Sweating efficiency
$$= \text{Rate of sweat evaporation} / \text{rate of sweat production}$$

Different areas of skin vary in their ability to produce sweat. Because of the variation in evaporating power of the local ambience, the sweating efficiency of the body is always lower than 100%, even when thermal balance is easily achieved. It decreases markedly, however, when this balance becomes difficult to maintain. This concept adds little to the understanding of the sweating mechanism. However, it is useful in applied thermal physiology.

The same reasons that explain why sweating efficiency is lower than 100% also explain a paradoxical fact. It is customary to believe that sweating differs markedly from one subject to another. However, we have seen above that the level of sweat secretion is strictly related to the need for thermal balance. This means that *two subjects, having the same physical morphology and the same metabolism, and being in the same environment, have approximately the same thermal requirements, that is, the same rate of sweat secretion.* Small differences, however, may occur. The most classical one is that observed between man and woman. It has been generally observed that women sweat less than men in a given hot environment. Moreover, women show a lower heat tolerance than men, based on tests measuring the time subjects could remain in a hot room.

Some facts can explain this sex difference. The lower basal metabolic rate of women enables them to maintain thermal equilibrium by sensory heat exchange at higher environmental temperatures than men. It has also been observed that women usually have lower skin temperatures at rest in a neutral environment. This difference is due to a different level of the cutaneous vasomotor tone. Such a difference could also be explained by changes in the spatial distribution of the sweat glands or in the time constant of the onset of sweating. However, experimental data are poor on this subject. On the other hand, Fox *et al.* (1969) have suggested that "women tend to be less acclimatized than men even when the two sexes live in the same climate and apparently follow a similar pattern of life." The lower level of heat acclimatization in women is the result of greater reliance on vasomotor control, with less frequent recourse to sweating, than in men. This consequently leads to a lower level of "training" of the sweating mechanism. The observations that an unusually high level of physical training in women abolishes the sex difference in sweating is in accord with the theories of Fox *et al.*

Nevertheless, because of these small differences, the amount of liquid sweat appearing on the skin of different subjects may vary. It is customary to use the term *sweating* only if sweat drops are seen on the skin. Yet this may be but a small part of the total sweat secretion. A subject can, therefore, sweat at a high rate even though his skin appears almost totally dry.

We have seen that the degree of wetness of the skin results from the ratio between the sweat rate and the ability of the environment to remove water vapor. To describe this, another concept has been introduced, called the *skin required wetness*. However, this term should be avoided, because it implies that skin wetness is a cause of sweat evaporation, whereas it is actually the result of this phenomenon.

7.3. HEAT PRODUCTION UNDER COLD CONDITIONS

When the rate of metabolic heat production becomes lower than the rate of heat removal by the environment, i.e., when

$$\dot{H}_m < \dot{H}_e + \dot{E}_b$$

the body tends first to counteract this imbalance by decreasing skin temperature. This response only slightly reduces heat loss. However, if the decrease remains insufficient to compensate for the body's heat production, a second regulatory mechanism is brought into play: an increase in metabolic heat production. To achieve a new heat balance, the rate of additional heat production $\Delta \dot{H}_m$ must equal the difference between $\dot{H}_m$ and $(\dot{H}_e + \dot{E}_b)$, i.e., the (negative) heat storage:

$$\dot{H}_m - \left(\dot{H}_e + \dot{E}_b \right) = \dot{S} = \Delta \dot{H}_m$$

The increase in heat production may be voluntary or reflex. Initially the subject performs muscular exercise. However, since the heat produced is related only to the intensity of the exercise, it does not exactly compensate for the heat loss; it may be lower or higher. On the other hand, exercise may change the value of the heat removal itself by modifying the convection coefficient through an increase in the relative air movement. This phenomenon is especially marked in water: Swimming in cold or even cool water markedly increases convection and therefore heat loss. For this reason heat production by voluntary muscular exercise is not a true thermoregulatory mechanism.

The only mechanism an organism can then employ is stress-adapted thermogenesis produced by a complex reflex mechanism. The response to cold is produced by modifying the metabolic reactions, leading to increased heat production. The most apparent response of this kind is the particular contraction of the muscles that we call *shivering*. However, in cold conditions thermogenesis may occur in other organs by a different mechanism; it is then called *nonshivering thermogenesis.*

7.3.1. Nonshivering Thermogenesis

During the last decades of the nineteenth century and the early part of this century, nonshivering thermogenesis was considered to be the major response to cold stress. It was assumed that this form of heat production arose from the viscera, especially the liver. In the following decades, the role of shivering has been seen to be increasingly important and nonshivering thermogenesis has come to be considered as only an auxiliary mechanism.

Yet the discovery of the role of *brown fat* or *brown adipose tissue* in heat production has awakened new interest in nonshivering thermogenesis. Brown fat is a special tissue found only in some subcutaneous regions, such as the scapularis. Its name comes from its particular color. It is absent in many species and is minimal in adults. In humans it is present in the neonate and disappears rapidly during growth. Histologically, it has two distinct characteristics. (1) Unlike normal adipose tissue, it is multilocular. (2) It is rich in sympathetic nervous fibers. Physiologically its metabolism is low, at least in a neutral and warm environment. In cold, however, its metabolism is greatly increased, under the control of the sympathetic nervous system and with a fairly rapid response. The thyroid gland is probably another controlling factor, but the thyroid response is delayed. (For a further discussion of the role of brown fat, see Section 7.3.3.)

7.3.2. Shivering Thermogenesis

Shivering is a particular kind of muscular contraction in which mechanical efficiency is very low. For this reason, the energy generated by shivering appears almost entirely in the form of thermal energy. The role of shivering in thermoregulation was recognized a century ago, when Richet claimed that "Shivering is the process of thermal regulation which produces heat by generalised contraction of muscles." Its origin is reflex, but it may be partly subjective and temporarily under the control of the will.

Muscular contractions during shivering are of short duration but are repeated, appearing as successive bursts. The intensity of shivering is related to the number of contracting muscles (spatial summation) and to the frequency of the contraction bursts (temporal summation). These contractions occur without movement of the limbs; this explains the low level of mechanical efficiency.

7.3.2.1. Methods for Investigating Shivering

The methods for studying shivering are numerous.

1. Observation by palpation of the muscle, even by the subject himself, can give information about the occurrence of shivering although less about its intensity. For this reason, this method is unsuitable for experimental study.

2. Since shivering is a mechanical phenomenon, it can be recorded by mechanical systems. It is therefore possible to use such devices as piezoelectric transducers, pressure sensors, and strain gauges.

3. The most widely used method is electromyography (EMG). Recordings can be made from cutaneous or, preferably, intramuscular electrodes. The latter technique provides a more precise localization of shivering muscles. Quantification of EMG is needed in order to establish the relationship of shivering to the electrical activity recorded. EMG amplitude is generally used for this purpose. However, a better result is derived from the integration of EMG impulses (integrated EMG).

4. Finally, since shivering appears to be an important means of thermogenesis, at least in man, direct measurement of the heat produced is a good means of evaluation. This may be measured by direct calorimetry. The simplest technique, however, is to measure oxygen consumption, assuming that the mechanical fraction of the energy used is close to zero.

7.3.2.2. Nature of Shivering

Even at rest, muscle has a contractile activity that constitutes the *muscular tone*. When the body is submitted to cold, the first response (except for the cardiovascular response) is an increase in this basal contractile activity to achieve what is called *preshivering tone*. When studied in a muscle bundle, it is evident in the extensor muscles before the flexors. It does not appear to differ markedly from physiological tone.

When true shivering appears, the electric activity of the muscle is characterized by an increase in the impulse frequency at the level of a given unit and especially by an increase in the number of discharging units (spatial summation). At the start of true shivering, the discharges of the various units are not synchronous. A certain delay is necessary to obtain the synchronization of activity of these units that leads to the well-known rhythmic contractions of shivering.

The frequency of shivering bursts is about 10–12 Hz. This frequency, and especially the amplitude of the bursts, may vary under certain influences. For instance, ventilatory activity modifies shivering: Its frequency and amplitude are increased during inspiration, but are decreased during expiration. On the other hand, voluntary muscular activity inhibits thermal shivering. This phenomenon might result from a direct inhibitory control of the motor center, or from a thermoregulatory process: the voluntary contraction of muscles producing heat, which suppresses the need for additional heat production. Although this last explanation is valid, inhibition exists even if voluntary movements are of short duration and weak amplitude, i.e., even if the quantity of heat produced by this voluntary activity is negligible. Such inhibition probably effects the motorneuron directly and is due to a reflex originating from the muscular stretch receptors. The relationship between the shortening of a muscle (when it contracts) and a decrease in shivering activity has been demonstrated in an inverse experiment: When a shivering muscle is stretched, its shivering activity is increased. Such a relationship can probably also explain the influence of ventilatory activity on shivering.

7.3.2.3. Localization of Shivering

Almost all muscles can shiver, including those of the tongue. The only exception is probably the eye muscles. Nevertheless, there are important differences in the ability of various muscles to shiver.

Generally, shivering begins in the muscles of the upper part of the body, including the masseters (the classical chattering of the teeth) and particularly the muscles of the scapular belt. From there, it spreads to the paravertebral muscles and to the lower limbs. Unlike pathological trembling states, thermal shivering always remains more intense in the proximal muscles than at the extremities.

7.3.2.4. Mechanism of Shivering

The peripheral generation of shivering is achieved in the spinal cord by the motoneurons. But the control of thermal shivering is the function of nerve structures called thermoregulatory centers, which are described in Chapter 6. However, numerous experimental observations show that the frequency of the discharges descending through the spinal cord to control shivering differs from the frequency of the shivering process itself. It may therefore be concluded that the generation of these oscillatory contractions must come partly from the activity of the spinal neurons or from a reflex phenomenon.

Sectioning the posterior spinal roots suppresses the sensible afferences and disturbs shivering markedly. This finding indicates that a reflex phenomenon and the control of the thermoregulatory centers combine to produce shivering. Numerous observations show that the *muscle* or *myotonic reflex* contributes to shivering, in addition to its normal role of maintaining muscular tone. This reflex finds its origin in the activity of the stretch receptors located in muscles. For a given length of muscle, a stretch receptor responds with a stable frequency of discharges, which can be recorded in its neuron fibers (tonic activity). When the receptor is stretched, the frequency of discharge increases markedly and reaches a peak before decreasing progressively toward the level corresponding to its new length (phasic activity). Taking account of the conduction delay along the neuron fibers, the combination of the two patterns of activity, with the phasic predominant, may result in the 10- to 15-Hz oscillations of shivering (Stuart *et al.*, 1963). However, this explanation is probably not totally satisfactory because these oscillations appear in muscles with differing fiber lengths. It may be that the muscle itself, through its own frequency of contraction, could act as a buffer, introducing new modifications of the discharge frequency it receives.

Shivering, then, appears as the result of a deep modulation acting on the muscle reflex under the control of the thermoregulatory centers. The pathway of this control, of course, travels along the spinal cord. Experimental transversal sectioning of the spinal cord suppresses the shivering ability of all muscles that are situated below the level of the section. A similar phenomenon is observed in paraplegic humans.

In both clinical and experimental situations a dissociation between voluntary movement and thermal shivering is observed. Therefore it is probable that the spinal pathways for the diencephalic control of shivering do not all pass along the pyramidal tract. There have been some discussions on the precise localization of these pathways, probably because of interspecies differences. In the dog, for instance, the integrity of the anterior spinal fasciculi seems necessary for transmitting the command for shivering. In man, however, cutting the spinothalamic fasciculi (to relieve severe pain) suppresses shivering only below the site of the section. In fact, control of shivering is thought to be a function of the reticulospinal fasciculus (which is close to the spinothalamic region).

7.3.2.5. Contributing and Inhibitory Actions

The system that controls shivering is complex, with many differing modes of action. These actions can operate at different levels and are triggered by nervous or humoral (or even pharmacological) impulses.

Since shivering corresponds to a particular modulation of muscle reflex, some actions that can affect this reflex can also influence shivering. For instance, the involvement of the γ system* is suggested by the fact that the electrical stimulation of some nervous structures enhances or inhibits both shivering and the γ system. It has been observed that subjects presenting with cerebellar lesions are found to have abnormal shivering reflexes. On the other hand, electrical stimulation of the cerebellum will generally inhibit shivering.

---

*The γ system is that system which allows the direct control of the muscular part of the stretch receptors by the central nervous system, principally the reticular formation.

The inhibitory reticular system has also been implicated in the control of shivering. One of the major influences modifying the activity of the reticular system is represented by the spinals of the sinocarotid nerves. The electrical or chemical (e.g., cyanide) stimulation of the central ends of these sinocarotid nerves induces a decrease in shivering activity.

Humoral actions are different and often complicated in their patterns of influence. They act generally on the thermoregulatory receptors or centers. The best-known influence is that of the anesthetic and depressive drugs. Some of these, like chlorpromazine, have an inhibitory effect on shivering, and are therefore used clinically in conjunction with exposure to cold to produce hypothermia. Others, on the other hand, enhance shivering (or decrease its threshold), as do the inhibitors of monoamine oxidase and the derivatives of diabenzazepin. By bringing about a significant increase in shivering and, consequently, increased heat production, some of these latter drugs are able to induce hyperthermia. In some subjects, such hyperthermia may be significant and rapid in onset; in anesthesiology it is called *malignant hyperthermia* (see Chapter 9).

We have also seen that other drugs act indirectly, by modifying the activity of the reticular system (e.g., inhibition of shivering by cyanide).

Shivering can be inhibited at the periphery, for example, by blocking the neuromuscular synapses with curare or its derivatives. This action appears with very small doses, lower than those that act on ventilation.

The actions of oxygen and carbon dioxide are more complex, because central and peripheral influences are intricate. An increase in blood Pco_2 is generally followed by a decrease of the intensity of shivering and the same result appears when Po_2 decreases. This fact probably explains the decrease of thermoregulation observed in a cold environment at high altitude.

Finally, the normal mediators of the nervous system are also involved in shivering, especially the catecholamines.

7.3.2.6. Catecholamines and Thermal Shivering

The secretion of catecholamines by the adrenal glands is markedly enhanced during the response to cold stress, and their levels in the blood consequently increase. On the other hand, injection of catecholamines rapidly induces an increase in metabolic heat production, adrenaline appearing to be more effective than noradrenaline. This increase in metabolism is more marked with high initial catecholamine levels.

A direct effect of catecholamines on tissues has been assumed because of the well-known action of these substances on some metabolic pathways. For example, adrenaline produces hyperglycemia by increasing hepatic and muscular glycolysis, and it also stimulates the release of free fatty acids from adipose tissue, thus mobilizing the lipid reserves. Nevertheless, this pattern of activity does not totally explain the increased thermogenesis due to catecholamine injection.

The injection of catecholamines produces more thermogenesis in an already shivering subject than in a subject under neutral conditions. Catecholamines may act not only on nonshivering thermogenesis, but also on shivering intensity. This action would stimulate the reticular system.

7.3.2.7. *The Importance of Cold-Induced Thermogenesis*

The amount of heat produced in response to cold is determined by the cold stress itself, increasing with falling ambient temperature. However, it is very difficult to determine when thermogenesis reaches its peak. During a burst of shivering, this maximal value has been found to be as high as 10 times basal thermal metabolism. However, because these bursts are successive, the mean peak value of heat production that a body can maintain for a period on the order of several hours is reduced. In man the mean peak value appears to be only 2–3 times the basal level, but it is higher in some other species, such as the dog. For this reason, the physiological capacity of the human body to resist cold stress is very poor, and thus man often resorts to behavioral (and technological) regulation.

7.3.3. Relative Importance of Shivering and Nonshivering Thermogenesis

In those species possessing brown fat tissue, shivering and nonshivering thermogenesis are used simultaneously, although at different rates. For example, when suddenly exposed to cold, the rat raises its thermal metabolism by shivering. This response can be triggered very rapidly. After several hours of exposure, however, shivering decreases, and nonshivering thermogenesis takes over and may even replace shivering completely.

The problem is different when brown fat is absent, as in adults. In man, for instance, shivering remains the most important source of increased thermogenesis. Even after several hours of cold exposure, it can represent at least 90% of the total heat produced to counteract the cold.

However, in the human neonate shivering is rare, even in a very cold environment. Unlike in the adult, 95% of the heat produced against cold by the neonate is generated without shivering, possibly in the brown fat—although evidence for this is indirect. The oxygen consumption of brown fat *in vivo* has been found to be about 60 $ml \cdot kg^{-1} \cdot min^{-1}$. Hull and Smales (1978) pointed out that "the amount of brown fat in a full-term infant of 3.5 kg required to achieve the increase in thermogenesis provoked by cold exposure at 25°C would be about 47 g." At birth, brown fat represents about 28% of the total fat content of the body, and this total represents 10–12% of body weight. In a neonate of 3.5 kg, about 100 g of brown fat would be found, an amount that is higher than the required value.

After birth, the role of nonshivering thermogenesis decreases rapidly and shivering becomes relatively effective. At 8 months, nonshivering thermogenesis accounts for only 30–40% of the total heat produced under cold conditions. At 1 year, nonshivering thermogenesis represents only 10%, whereas shivering accounts for 90%, i.e., the same figures found in the adult.

8

Acclimatization to Heat and Cold

The problem of the acclimatization of man to heat and cold is an ancient one. However, it has attracted the most attention in the years during and since World War II for military and economic reasons. Two questions arise: First, is a subject already living in an extreme climate more efficient than a subject coming to it from a temperate climate (natural acclimatization)? Second, if acclimatization exists, is it possible to produce it experimentally? This would prepare a subject from a temperate climate to adapt immediately on arrival (artificial acclimatization). Before answering these questions the concept of acclimatization and the word itself must be defined.

8.1. THE CONCEPT OF ACCLIMATIZATION

When a working man is rapidly subjected to a hot climate, he exhibits a progressive increase in his central temperature and his ability to work is impaired. After a few days, his ability for work increases but his central temperature tends to fall, although sweating increases. We say that the man has become *acclimatized*. In terms of central temperature and ability to work, acclimatization appears to be a benefit; however, the increase in sweating makes the subject more dependent on his environment, i.e., water supply. On the other hand, the camel, which is of course well adapted to heat, sweats very little but has a high central temperature. There is, therefore, more than one way to obtain optimum ability to work in in a warm environment. What appears to be good for one species is not necessarily good for another.

In terms of the physiological stress resulting from external stress the problem is also complex. Is the stress of increased sweating more or less than that caused by an increase in central temperature? In the camel an increase in core temperature produces little stress; in man it produces significant stress. This may be because hyperthermia impairs psychological function more rapidly than muscular function.

8.2. DEFINITION OF ACCLIMATIZATION

We have so far referred to two concepts: *acclimatization* and *adaptation*. Although often used interchangeably, they are not synonymous. The following definitions have been proposed by Goldsmith (1977):

☐ *Acclimatization:* "Those changes in the responses of an organism produced by continuous alterations in the environment."

☐ *Acclimation:* "Those changes in the responses of an organism produced by alterations within the lifetime of the organism."

☐ *Adaptation:* "Those changes occurring during a period of several generations."

These definitions are interesting because they highlight the effect of exposure time. However, it is necessary to add the concepts of *natural* and *artificial* acclimatization to these definitions.

On the other hand, the "Glossary of Terms for Thermal Physiology" (Bligh and Johnson, 1973) always uses *adaptation* to refer to genetic changes but uses the terms *acclimatization* for natural acclimatization and *acclimation* for artificial acclimatization:

☐ *Acclimation:* "A physiological change, occurring within the lifetime of an organism, which reduces the strain caused by experimentally induced stressful changes in particular climatic factors."

□ *Acclimatization:* "A physiological change occurring within the life-time of an organism which reduces the strain caused by stressful changes in the natural climate."

These last definitions point out the possible differences which may exist between natural and artificial acclimatization. They also point out the concept of thermal stress. However, the effects of thermal stress may prove ambiguous and may vary with the type of acclimatization.

These definitions ignore the role of man's behavioral ability to change his environment by means of technology, or even the ability of his neuropsychological functions to modify his perception of a stimulus. For instance, the pain from immersing the hand in cold water may progressively disappear if the subject can be persuaded to expose himself to this uncomfortable stimulus sufficiently often. It has been proposed that the term *habituation* should be used for this phenomenon (Goldsmith, 1977).

It appears that too accurate a definition of acclimatization is not appropriate. It is best that we limit our definition of *acclimatization* to the physiological changes that are due to artificial or natural climate and that produce an "enhanced ability to maintain functions [physiological and/or psychological] in the face of climatic stresses" (Budd, 1974).

8.3 NATURAL ACCLIMATIZATION TO HEAT

If an individual, accustomed to performing a heavy task with normal efficiency in a neutral or cool environment, is transferred rapidly to a hot environment, he becomes unable to maintain that efficiency. He is overtaken by lassitude, and may present nausea and even collapse, while his central temperature increases to abnormal proportions. However, after a few days, physiological adjustments take place, discomfort decreases, and the subject shows improved efficiency. His central temperature tends to decrease but remains, nevertheless, at a higher level than in neutral or cool conditions. The subject is said to be acclimatized.

The major physiological phenomena of acclimatization are the increase in sweating and the decrease in the latency of its onset. The

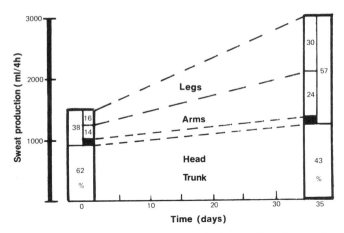

Figure 8.1. The increase in sweat production during acclimatization is greatest in the extremities.

increase in activity of the sweat glands during acclimatization is greater at the extremities than in the trunk (Figure 8.1). The first result of this total increase is to maintain heat storage at a lower value than at first exposure. Subsequently, the temperature rise tends to slow down, and heart rate falls slightly. Therefore, the thermal stress of the body, expressed as the level of central temperature and heart rate, is less than it was before acclimatization. On the other hand, the acclimatized subject, on average, drinks more than an unacclimatized subject.

8.4. ARTIFICIAL ACCLIMATIZATION TO HEAT

Artificial acclimatization is possible; in other words, the physiological changes observed during natural acclimatization can be induced by experimental conditions. Artificial acclimatization has provided a better knowledge of the process of acclimatization itself.

There are different protocols for obtaining artificial acclimatization, but all try to submit the subject to a large heat stress in order to

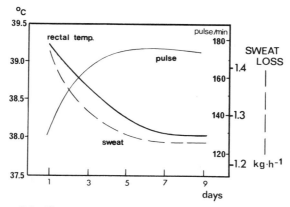

Figure 8.2. Changes induced by artificial acclimatization to heat.

induce an increase in central temperature and the appropriate regulatory responses of the organism. Since muscular exercise produces an increase in thermogenesis, acclimatization can be achieved when a physical task (e.g., walking, exercising on an ergometer) is performed in a hot environment for 1–2 h and is repeated for 8–10 days. Generally, ambient temperature is fairly low the first day and is increased on the following days. The results appear to be similar to those found with natural acclimatization (Figure 8.2). This technique has been applied intensively on military and industrial workers. For instance, Wyndham *et al.* (1965) used this technique to acclimatize 250,000 people working in the gold mines of South Africa.

However, acclimatization obtained by this method is more or less specific. If the subject has to carry out a different task, the artificially achieved acclimatization may not entirely transfer. It has therefore been suggested that heat acclimatization should not be confused with physical training. Furthermore, heat acclimatization would appear to relate more to the duration of the overall stress, i.e., to the increase in central temperature. For this reason, some authors, particularly Fox *et al.* (1963), developed a method in which hyperthermia is obtained without active work and can be accurately monitored (controlled hyperthermia).

An increase in central temperature can be due to an impairment of the evaporative mechanisms. Goldsmith (1977) achieved this in a protocol in which "the subject wears an all enveloping vapour barrier suit and lies on a bed. Body temperature is raised and controlled by covering the subject with an insulating layer and circulating hot air around him." The rate and temperature of this circulating air can be modified to control body temperature. The test is repeated over 10–20 days. Sweating increases linearly, session by session, in a fashion similar to that observed with the other techniques for artificial acclimatization.

The performance of men acclimatized by these techniques has been tested by making them work in heat before and after acclimatization. It has been concluded that acclimatizations achieved using controlled hyperthermia and "work-in-heat" techniques are similar. Yet despite the similarity in results, the mechanisms by which controlled hyperthermia induces acclimatization may differ from those involved in "work-in-heat" routines. It has been suggested that the increase in sweating in the latter case is due to two different mechanisms, one central (increasing the control of the hypothalamic centers on the sweat glands) and the other peripheral (by direct conditioning of the glands themselves). Recent evidence suggests that the peripheral mechanism should be more significant than the central mechanism.

On the other hand, if sweat cannot be totally evaporated, the skin becomes increasingly moist. Overhydration of the skin or prolonged hydration causes a secondary progressive decrease of sweat rate, called *hidromeiosis* (Henane, 1972).

8.5. ACCLIMATIZATION TO COLD

The question of whether true acclimatization to cold exists in man is not a simple one.

Acclimatization to cold is observed in some species of mammals. For instance, if rats are kept at 6°C for a month, they exhibit intense shivering accompanied by behavioral disturbances during the first days of exposure. Following this period, shivering decreases because the animals develop nonshivering thermogenesis. If, at that time, they are put in a colder environment, for example one at −18°C, they are able to maintain their central temperature at a tolerable level, whereas unacclimatized rats rapidly develop fatal hypothermia. This possibility for acclimatization in animals appears to result from an activation of the brown adipose tissue under the influence of increased catecholamine output by the adrenomedullary glands.

In man, however, the problem of cold acclimatization arises in other ways. One possible means of studying it is to observe people normally exposed to cold.

Surprisingly, cold acclimatization in *not* found in Eskimos. Their life conditions are such that only the face is exposed to cold air. It has never been possible to demonstrate nonshivering thermogenesis in these people.

Conversely, possible acclimatization has been observed in such primitive people as the Australian aborigines and the bushmen of the Kalahari desert, who live nude in a relatively cold climate and are submitted to cold stress, mainly during the night. When compared to Caucasians, it appears that they shiver less rapidly when exposed to the same cold stimulus. But there is no increase in metabolic rate, as is observed in animals. The decreased shivering is accompanied by a more marked fall in their central temperature.

A better tolerance to hypothermia has been observed by Japanese authors studying Ainu skin divers.

Attempts to produce artificial cold acclimatization in man have been made, using repeated or prolonged exposure of the subjects to cold stress. Although it has been shown that the thermogenic effects of injected noradrenaline in man would be potentiated by a period of cold exposure, the majority of experimental studies indicate that prolonged exposure to cold does not produce changes in basal metabolic rate. It appears, therefore, that man does not achieve general acclimatization to cold by developing nonshivering thermo-genesis. However, two other biological processes may play a role in increasing the resistance of man to cold: a change in peripheral insulation and peripheral vascular adaptation.

8.5.1. Change in Peripheral Insulation

Studies made during polar expeditions have revealed that man can more effectively maintain central temperature in a cold environment, without increasing metabolic rate, by laying down a thicker layer of insulating subcutaneous fat. Experimentation on the responses to a standard cold stress both before and after a long stay in a cold climate has led to the general conclusion "that rectal temperature falls less in tests after acclimatisation, while oxygen consumption does not rise to higher levels than in the test before acclimatisation" (Goldsmith, 1977). These conclusions agree with the assumption of better body insulation owing to acclimatization and also confirm the absence of increased metabolic rate after repeated exposure to cold.

8.5.2. Vascular Acclimatization

Contrary to heat stress, which affects the whole body and tends to produce hyperthermia (the initial cause of acclimatization), cold stress generally acts only on the skin and especially that of the face and hands. It has therefore been postulated that, if some degree of cold acclimatization can be expected, it should be obtained by applying the cold stimulus to these regions.

The habitual response to the immersion of a hand in cold water at 4°C (cold test) is an increase in sympathetic activity, marked by vasoconstriction, increased heart rate, and a rise in arterial blood pressure. Vasoconstriction contributes to a decrease in heat loss from the body, but the increased heart rate and blood arterial pressure

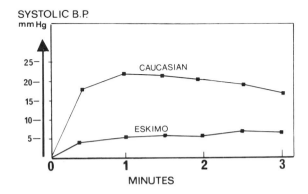

Figure 8.3. Increase in systolic blood pressure induced by a cold test in Caucasian and Eskimo [derived from Leblanc].

impose a stress situation. However, if Eskimos—whose face and hands are generally more exposed to cold than is the remainder of the body—and nonacclimatized Caucasians are submitted to this test, they respond differently. The former exhibit less sympathetic reactivity, and especially a diminished increase in arterial blood pressure (Figure 8.3). A similar observation has been made experimentally. Repeated exposure of the hand to a cold test over 10–20 days can induce a decrease in sympathetic reactivity, i.e., a smaller increment in heart rate and arterial blood pressure. However, there is little evidence that the face itself can acclimatize.

On the other hand, another phenomenon may be involved in local acclimatization. When the hand is plunged into ice water, one observes local vasodilation or reactive hyperemia, sometimes described as *paradoxical vasodilation* (see Chapter 5). It has been observed that those who habitually expose their hands to such a cold stress, such as fishermen, show a more rapid onset and a higher level of vasodilation, when their hands are put into ice water, than normal subjects.

In conclusion, it appears that, unlike other mammals, man is not able to acclimatize spontaneously or artificially to cold by increasing nonshivering thermogenesis. However, he may develop a better natural insulation by subcutaneous fatty tissues and be able to diminish the sympathetic strain induced by cold stress. Of course, the most effective "acclimatization" of man is his use of technology to suppress the stress of a "hostile" climate.

Disorders of Thermoregulation

The pathology of thermoregulation can be studied from two viewpoints: (1) Do the regulatory responses disturb other functions of the organism? (2) Is thermal function itself abnormal?

9.1. UNFAVORABLE EFFECTS OF THERMAL RESPONSES

9.1.1. Responses to Cold

The principal responses to cold stress are vasoconstriction of skin vessels and an increase in metabolic heat production. Constriction of the vessels of the skin may act unfavorably by decreasing the blood supply to the extremities and, therefore, by increasing the possibility of injury by frostbite. However, in a cool environment this is not possible, and the only effect would be an aggravation of the problem produced by arteritis (inflammation of the arteries). There are, however, few data on this subject.

Vasodilation of the peripheral vessels may be accompanied by a reflex vasoconstriction of the vessels in other organs. For example, digestive problems experienced in very cold weather can be attributed to vasoconstriction of the mesenteric vascular bed.

It has been reported that periods of intense atmospheric cold can increase the incidence of angina pectoris. However, even in this case the results are inconsistent. On the other hand, the increase in metabolic heat production created by shivering is generally insufficient to strain the cardiovascular system.

9.1.2. Responses to Heat

Thermal responses to heat can act unfavorably on other functions. This problem is a more widely recognized phenomenon. The main response to heat stress is perspiration, which induces water loss and, to a lesser degree, a loss of electrolytes, mainly NaCl. Body dehydration may be, by itself, a pathological phenomenon, especially in the infant. However, even if it remains at a moderate level, dehydration produces modifications in the blood, e.g., an increase in the hematocrit. There is, however, no substantial evidence for the possible pathological effects of this phenomenon.

Another response to heat stress is an increase in cardiac performance. Therefore, the stress from a hot environment would act unfavorably on patients in creating alterations in cardiac function. It has been observed that the number of deaths from cardiac and especially coronary diseases increases during the hot periods of the year. Men working at jobs in which they are continually exposed to hot environments also seem to have a greater mortality from cardiac failure. Recent statistical studies have confirmed and clarified earlier observations on the role of climate in heart disease. States (1977), for example, has compared the mortality from cardiac diseases during summer in two regions of the United States, one in a temperate climate (Pennsylvania) and another in a subtropical climate (Alabama). Paradoxically, he found mortality to be higher in the temperate climate. However, the study of mortality month-by-month shows that there is a peak during summer. Mortality also correlates with temperature *variation* better than with *mean* temperature: In subtropical climates, mean temperature is moderately high, but with less fluctuation. This agrees with other observations made during major heat waves. It has been observed that mortality during heat waves was greater in regions that are normally temperate or even cool. During severe heat spells in the USA during July 1966, mortality increased by 36% (above the normal predicted value) in New York and by 56% in St. Louis. The maximum mortality rate was observed on the day following the highest recorded temperature.

It has repeatedly been observed that the correlation between weather characteristics and death increases with age. One possible explanation for the observation that mortality depends on temperature variation rather than the actual level of temperature is acclimatization. Subjects living in a subtropical climate are acclimatized to heat. Therefore, when a major heat stress occurs, as during a heat wave, it is better tolerated by them than by those who normally live in a temperate climate and who are therefore not acclimatized.

9.2. ABNORMALITIES IN THERMOREGULATION

Internal body temperature can fall outside the normal range by two distinct processes:

1. The regulatory mechanisms operate normally, although at their maximum they are inadequate to counterbalance thermal stress. The body is then subjected to a thermal force. This may be called "passive" hyper- or hypothermia.
2. Abnormalities in regulatory mechanisms occur, which lead to a failure to adjust the response required for thermal balance, either internal (thermal metabolism) or external (modifications of environmental conditions). Fever is one example of a failure of the body to adjust its thermal balance correctly. However, the body retains the ability to sense and react to changes in external stress.

Occasionally, these two situations may interact.

9.2.1. Overshoot in Thermoregulatory Mechanisms: "Passive Abnormalities"

These abnormalities may lead to either hypo- or hyperthermia according to the type of thermal imbalance.

9.2.1.1. Hypothermia

Hypothermia can be observed in healthy subjects who are subjected to an intense cold stress. Although there is a regulatory increase in metabolism, heat loss (depending on clothing) remains higher than heat production. Therefore, negative heat storage occurs and increases with time, while internal temperature progressively falls. This is termed *accidental hypothermia.*

Accidental hypothermia can be observed in arctic regions and in mountains (hypoxia increases the rate of onset of hypothermia). However, it occurs more frequently during water immersion. As already shown, water has a convective power 25 times greater than that of air. Heat removal in an immersed subject can be very high and rapid, especially if there is motion between the subject and the water, e.g., in a swimming subject. The importance of heat loss is such that immersion in still water at 15°C can be lethal (Figure 9.1).

Since heat loss is related not only to temperature but also to such other factors as the relative movement with respect to a fluid, it is better to describe the symptoms of hypothermia in relation to internal temperature (Figure 9.2). The triggering of regulatory mechanisms is fairly rapid and does vary according to the subject, but generally appears when the central (rectal) temperature falls to 36.5°C. The first response, other than cold sensation, is the onset of shivering, which rapidly becomes involuntary. The first manifestations to appear are mainly psycological: At about 35°, reasoning slows down and the subject lets himself go and refuses to do

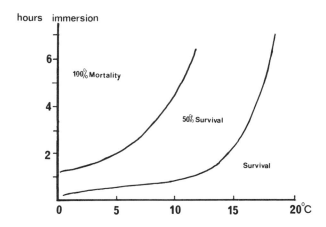

Figure 9.1. Estimated risk of mortality by immersion in cold water [from Molnar, 1946].

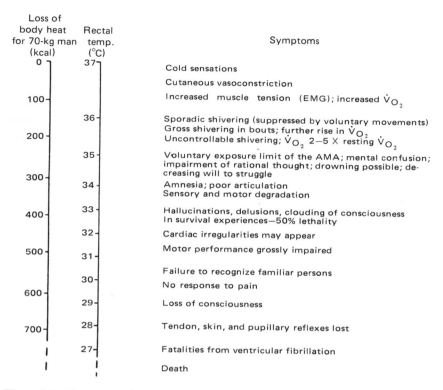

Loss of body heat for 70-kg man (kcal)	Rectal temp. (°C)	Symptoms
0	37	Cold sensations
		Cutaneous vasoconstriction
100		Increased muscle tension (EMG); increased $\dot{V}_{O_2}$
	36	Sporadic shivering (suppressed by voluntary movements)
200		Gross shivering in bouts; further rise in $\dot{V}_{O_2}$
		Uncontrollable shivering; $\dot{V}_{O_2}$ 2–5 X resting $\dot{V}_{O_2}$
	35	Voluntary exposure limit of the AMA; mental confusion; impairment of rational thought; drowning possible; decreasing will to struggle
300	34	Amnesia; poor articulation
		Sensory and motor degradation
400	33	Hallucinations, delusions, clouding of consciousness
		In survival experiences—50% lethality
	32	Cardiac irregularities may appear
500	31	Motor performance grossly impaired
	30	Failure to recognize familiar persons
600		No response to pain
	29	Loss of consciousness
700	28	Tendon, skin, and pupillary reflexes lost
	27	Fatalities from ventricular fibrillation
		Death

Figure 9.2. The progressive failure of body systems owing to heat loss [from Webb, 1976].

anything that could decrease the cold stress. By this time, heat loss is about 1000 kJ (~ 250 kcal). The value of 35°C for the central temperature appears to be the extreme limit of voluntary exposure in well-trained subjects, such as deep sea divers. It also sets a limit for volunteer subjects in cooling experiments.

When a central temperature of 34°C is reached condition worsens and speech becomes less fluent. Vision is obscured and there is a net decrease in muscular activity. At 32°C the heart exhibits rhythmic abnormalities and impulse conduction is affected. Consciousness is lost between 30° and 29°C and death by ventricular fibrillation occurs on the average at 27°C (see Figure 9.2). At this time, the loss of body heat is about 3000 kJ (~ 700 kcal).

Webb (1976) has pointed out that the practical determination of fatal heat loss is difficult. It requires the measurement of central and mean skin temperature, but the proposed weighting coefficients seem to be unadaptable to low temperatures. The temperature of the auditory canal, sometimes used for measuring central temperature, is now considered invalid for this purpose, because it is partly under the influence of cutaneous blood flow. Rectal temperature is more reliable than esophageal for this purpose.

9.2.1.2. Hyperthermia

Hyperthermia may be defined as an increase in central temperature. Physically speaking, it results from an excess of heat gains over heat losses. As already defined, hyperthermia could be called "passive" when the imbalance results from an overshooting of the regulatory limits and "active" when it results from an abnormal response by control mechanisms.

Unlike hypothermia, which appears only in a specific condition, the physiological causes of passive hyperthermia are multiple. They are generally classified as *accidental hyperthermia*, as in the following examples:

1. The subject produces a normal (basal) amount of heat, but is in an extremely hot environment, and/or is subjected to a strong radiant flux (e.g., exposed to the sun in hot climate). Although sweat flow is maximum and evaporation is readily attained, evaporative heat loss cannot compensate for the total heat gain.
2. The subject produces a significant amount of heat that cannot be totally lost even though the environmental temperature is not excessive. This situation is encountered during intense muscular exercise in a temperate climate (e.g., with marathon runners and racing cyclists).
3. The mechanism by which a subject can lose heat is impaired. This can occur under two conditions:
   a. When the subject is unable to attain an adequate sweating rate owing to dehydration (in this case, the term *passive hyperthermia* is not totally correct).

b. The environment has a decreased evaporative power because it is warm and very humid.
4. The preceding conditions can occur together, as in the case of a subject undergoing intense muscular activity in a hot and humid climate, without the possibility of replacing his water loss.
5. Finally, there may be another condition in which sweating corresponds not only to a loss of water, but also to a loss of electrolytes, mainly NaCl. If a subject is maintaining his thermal balance by sweating (and evaporation) his demands can only be to replace his water loss, but not his loss of electrolytes. In this case he may exhibit other symptoms of thermal distress.

The first signs of hyperthermia are mainly neuromuscular and psychological: fatigue, apathy, decreasing will to make an effort, and ultimately, hallucination. However, because of the multiple causes of hyperthermia, the clinical results may vary.

Heat fatigue (*syncope*) can occur in moderate hyperthermia and without major sweat loss in the unacclimatized subject. This will probably result in a strong cutaneous vasodilation and therefore a shift in blood distribution. *Heat stroke* results from increased heat storage, which cannot be compensated for by evaporative heat loss. Two forms may be distinguished. In one the rate of accumulation may be excessive, as in a subject exercising in a hot and sunny climate. In the other both heat gain and internal temperature increase rapidly before the regulatory mechanisms, particularly sweating, can be triggered or even become adequate. In this case, the symptoms of hyperthermia appear suddenly in the absence of marked sweating. Death can occur when central temperature reaches 42°–43°C, thus affecting the nervous system and/or causing cardiac failure.

If the rate of accumulation is slightly reduced, yet remains high, sweating is triggered. However, even at its maximal rate, it cannot be sufficient to compensate for the high rate of heat gain. The onset of hyperthermia is less rapid but heat disorders progressively appear. In this case, the symptoms of hyperthermia are accompanied by intense sweating. Death can also occur by cardiovascular collapse, but at a lower central temperature than in the preceding case. Therefore, the presence of sweating during heat stroke more or less depends on the speed at which heat builds up and on the rate at which the sweat glands can react. If the loss of water and electrolytes is compensated for by drinking, but heat gain continues to exceed heat loss, hyperthermia will also progressively increase and may be lethal. Some authors use the terms *heat hyperpyrexia* to characterize this increasing central temperature with a sustained high sweating rate.

Heat exhaustion is related to the situation in which the subject cannot replace the water and electrolytes lost by perspiration. The disruption of the thermal balance is therefore due to a fall in evaporative rate. The major signs are thirst followed by hallucinations and delirium. Death occurs by both hyperthermia and cardiovascular collapse. If the subject has the possibility of even partly replacing the water lost by sweating, but not the electrolytes, hyperthermia can still occur, but the clinical signs may be nausea, vomiting, and muscle cramps. Cardiovascular failure can subsequently develop.

9.2.2. "Active" Impairment of Thermal Balance

Active hypo- or hyperthermia may result from the abnormal functioning of thermal control. This may, for example, induce an increase in body heat production that is not related to the actual amount of heat lost. In this case, hyperthermia occurs and is then secondary to the control failure.

However, control failure is not the only cause of hyperthermia. Particular environmental conditions must also be present for the hypo- or hyperthermia to occur. Consider, for example, the case in which thermal control has been destroyed or pharmacologically inhibited. If metabolic heat production remains constant, central temperature will increase or decrease according to the extent that heat is removed by the environment. If, by chance, heat removal equals heat production, central temperature will remain constant. Therefore, total or partial failure of the body's thermal control may not necessarily result in a change in central temperature. The best example of this case is temperature disturbances produced by pharmacologic agents, such as alcohol or certain anesthetics. Alcohol has a peripheral effect by depressing vasomotor tone, leading to increased heat loss. But it also depresses the ability of the body to increase its heat production. For this reason, a subject who is alcoholically intoxicated will exhibit hypothermia. The degree of this hypothermia will be related to the ambient temperature: The lower the temperature, the more marked will be the hypothermia. If the subject is placed in a slightly warm environment, hypothermia will not occur.

An example of hyperthermia caused by a disturbance of the central control is observed in anesthesia and is known as *malignant hyperthermia*. In rare cases, some subjects during anesthesia develop an increase in central temperature. This rise may be very rapid and leads to death by hyperthermia. This increase is produced by an abnormal response of the central control that triggers the effector responses to hypothermia, including muscle contraction and shivering, without physical or physiological reason, while mechanisms of resistance to hyperthermia, like sweating, are depressed. Although the increase in metabolic heat production may be moderate, central temperature rises owing to the lack of sweating.

However, pathophysiological causes are not solely responsible; environmental conditions also play a role. The environment of the anesthetized subject (e.g., the high temperature of the room, radiating sources, bedclothes) contributes to a low rate of heat removal. Furthermore, when central temperature is monitored by a rectal probe, the thermal inertia of this region means that rectal temperature variation is slower than the rate of central temperature increase. The degree of hyperthermia may therefore be underestimated.

9.3. FEVER

Fever is the best-known example of hyperthermia caused by a disturbance in the thermoregulatory mechanisms. It is produced by abnormal *pyrogenic* substances in the blood that are released when the body is submitted to bacterial infections or inflammation of nonbacterial origin. The terms *exogenous* and *endogenous pyrogens* have been used to divide the resulting fevers into those produced by external agents as bacteria and those observed during the development of carcinoma. The term *endogenous pyrogen* is now exclusively applied to the pyrogen produced by the leukocytes.

9.3.1. Bacterial Pyrogens

The most active bacterial pyrogens are the endotoxins produced by gram-negative bacteria. They have a high molecular weight. Their pyrogenic activity is very high, for an intravenous dose of $1 \text{ ng} \cdot \text{kg}^{-1}$ body weight is enough to induce a pyrogenic response in rabbits.

The most active part of these microbial agents is formed by the liposaccharides of the cell wall. It is thought that calcium may be involved in the cross-linkage of polysaccharide subunits. Although synthetic polysaccharides may have a pyrogenic activity, other molecules are involved in the formation of specific poisons known as endotoxins, particularly the phospholipids and certain proteins.

On the other hand, some chemical compounds that differ from typical endotoxins can also show pyrogenic activity, e.g., dextran, inorganic colloids, viruses and gram-positive bacteria. When endotoxins are injected repeatedly, the body develops a tolerance, via an immunologic mechanism.

9.3.2. Endogenous Pyrogens

It has been recognized that the common route of pyrogenic activity is the release of a pyrogenic substance called endogenous pyrogen (EP) by leukocytes, under stimulation by bacterial pyrogens and other pyrogenic substances. EP differs from bacterial endotoxins in that it is more labile and is even able to induce fever in animals made resistant to bacterial pyrogens by repeat injections.

The release of EP may be brought about by the action of the bacterial endotoxins in the blood, by substances resulting from tissue alteration, or by synthetic substances such as polysaccharides or polyribonucleotides. This explains the delay (usually about 2 h) that is observed between the time of an exogenous pyrogen injection and the onset of fever, whereas injection of EP produces fever very rapidly.

EP may be also produced *in vitro* by the incubation of bacterial endotoxin with whole blood. The amount of EP produced is directly related to the number of leukocytes incubated with the exogenous agent.

Blood cells do not contain pyrogen before activation. Production and release of EP first requires an interaction between the stimulating agent and a cellular precursor. Penetration of the stimulating agent into the cell is probably achieved by phagocytes. The synthesis of EP requires a delay of 1–2 h. It may be that this time is needed not to produce EP from the precursor but to form an enzyme involved in the transformation. The presence of RNA is necessary but inhibitors of protein synthesis (e.g., as puromycin, actinomycin D) can prevent the formation of EP.

The release of EP occurs from 2 to 6 h after phagocytosis, when the intracellular levels have reached a high value. The release of EP probably becomes self-limiting in preventing further release. A new burst of production and release can nevertheless occur after a short delay.

The chemical composition of EP—beyond the fact that it is a small protein—is not yet totally known.

9.3.3. Mode of Action of Endogenous Pyrogen

The main site of action of EP is the preoptic area of the anterior hypothalamus (AH-PoA). The injection of very small amounts of EP into this area or into the third ventricle is rapidly followed by an increase in central temperature, but no such effect is observed if the injection is made into other brain areas.

It has been suggested that monoamines like noradrenaline or 5-hydroxytryptamine (5HT) would mediate the hyperthermic effect of EP because these substances are known to produce changes in T_d when injected into AH-PoA. However, these substances act differently in different species. For instance, in cats 5HT raises T_d and noradrenaline lowers it, whereas the opposite effects are obtained in rabbits. Moreover, depletion of hypothalamic monoamine content does not inhibit the production of fever following an injection of EP. Therefore, fever appears not to be primarily dependent on monoaminergic pathways.

Although there is some reason to suspect that EP-induced fever might involve the cholinergic pathway, the main mediators are probably the prostaglandins of the E series (PGE). Prostaglandins are the normal constituents of hypothalamic tissues, and their role in the mediation of febrile response has been suggested by many experiments.

The injection of either PGE_1 or PGE_2 intracerebrally in very small amounts is rapidly followed by an increase in T_d. The latency of this febrile response is less than that observed following EP given in the same area. Destruction of the entire AH-PoA makes animals unable to regulate their T_d. PGE_1 injected directly into the damaged AH-PoA no longer produces a temperature response. On the other hand, antipyretic drugs inhibit the synthetase that elaborates prostaglandins from their precursor. For these reasons, it has been proposed that prostaglandins may be released by leukocyte pyrogen in the hypothalamus, and that these fatty acids may mediate fever response to pyrogens.

However, the prostaglandin hypothesis has been recently subjected to criticism. For instance, continuous infusion of EP produces a steady-state fever in rabbits. When salicylate, which is a poor antipyretic drug in this species, in infused at the same time as the pyrogen, there is no rise in PGE concentration, but the fever remains unchanged. On the other hand, the prostaglandins are formed from a precursor, arachidonic acid. During this transformation, other compounds, called thromboxanes, are also created. Injection (in cats, rats, and rabbits) of arachidonic acid also produces fever, which is, furthermore, blocked by antipyretic drugs such as indomethacin. But a dose of this inhibitory substance sufficient to abolish a PGE fever has very little effect on arachidonic acid fever. For Hellon (1970), this implied that the thromboxanes, rather than PGE, may be responsible for generating fever, and he noted that "the testing of this hypothesis will give us new insights into the complex processes involved in the mediation of fever."

9.3.4. Role of Sodium and Calcium Ions

Feldberg (1971) and Veale and Cooper (1975) have shown that the constancy of temperature may depend on the balance of Na^+ and Ca^{2+} in the hypothalamus. Perfusion of the third ventricle (in the cat) with a 0.9% NaCl solution produced a vigorous shiver and a subsequent increase in T_d. Addition of $CaCl_2$ prevented this hyperthermic action. It is suggested that Ca^{2+} acts as a "brake" that prevents Na^+ from exercising its hyperthermic effect. This explanation is not in contradiction to the probable role of endogenous pyrogen. One conciliatory assumption is that either EP or PGE, or thromboxanes, may act by removing the Ca^{2+} brake, allowing Na^+ to exert its hyperthermic actions.

9.3.5. A Possible Mechanism for Fever Production

Figure 9.3 illustrates a possible mechanism for fever production.

In terms of thermoregulation, fever is produced by a disturbance of the thermal balance. After injection of a single dose of pyrogen two successive stages occur: first, marked shivering, which induces an increase in heat production and a subsequent rise in central temperature, and second, profuse sweating with reduction in temperature to normal. Fever is thus produced by an abnormal triggering of the mechanisms normally used against a cold stress to restore the thermal balance.

When the cause of the fever (the pyrogen) has disappeared, thermoregulatory mechanisms react against the abnormal heat storage by bringing into action a heat loss mechanism, i.e., sweating. It appears that only the first part of the hyperthermic burst, i.e., the increase in heat production, represents abnormal regulation. It is questionable whether the centers trigger an abnormal response that is unnecessary for thermal balance. There are two possible answers: Either the centers are impaired and have an inadequate response, or they are normal but receive false signals about the thermal state of the body. In the first case the primary cause of the abnormality must be in the centers themselves; in the second case it must be in the thermoreceptor or interneurons.

The first theory for fever was proposed more than a century ago by Leibermeister (1875). He described fever as a change in the reference level (set point) of body temperature. This theory is commonly used but more as an academic than a scientific explanation. It has previously been emphasized (Chapter 6) that the thermoregulatory system is too complex to be explained by a simple set point model. Recent experimental findings show that the second theory is

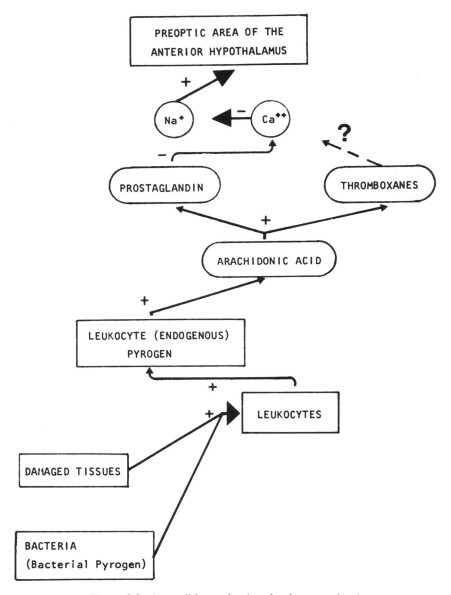

Figure 9.3. A possible mechanism for fever production.

probably correct. Some authors [Cabanac *et al.* (1968), Eisenmann (1972), Hellon (1970), and others] have demonstrated "that the properties of the temperature-sensitive neurons in the AH-PoA are modified when a pyrogen is systematically given. The neurons excited by hypothalamic warming had a decreased temperature sensitivity after pyrogen while the cold reacting cells became more temperature-sensitive. Temperature insensitive neurons were not affected" (Hellon, 1970). It appears therefore that fever is due to a nonadapted response of the centers. This is secondary to the false response of these centers by the central thermosensors, which are affected by the pyrogen. When pyrogen disappears from the blood these sensors recover their normal function. The centers receiving correct information now trigger mechanisms to eliminate the stored heat that had accumulated during the disturbance. The physiological and physical facts are now clear enough for us to assume that a change in the set point is not necessary. At best the set point hypothesis remains of purely academic interest.

9.4. THERMOGRAPHY IN DISEASE

Within the context of this book detailed descriptions of the clinical uses of thermography have been omitted. Clinical opinion and acceptance of thermography are still changing. For many clinicians, body temperature is mistrusted as a diagnostic tool, apart from basic oral measurements. Oral temperature, being an approximation to core temperature, has a limited range, and is only significantly raised or lowered in extreme clinical states.

9.4.1. Applications

The concept of skin temperature is complex. Nevertheless certain patterns of temperature distribution occur in the average subject who is free from disease. A number of pathological conditions do modify skin temperature distribution in some way. When these modifications are large enough, or very localized, thermography can be a useful method of investigation. The effect of the disease may present thermal changes in one of three ways. First, the pattern may be unchanged but the temperature range may be grossly abnormal. Second, the pattern of temperature distribution may be modified from the accepted norm. Third, the change may only be evident, or be exaggerated, in response to stress. (This may take many forms, but includes the use of drugs as well as mechanical and thermal stress.) The advantages of noninvasive measurement mean that thermography can be used successfully to monitor disease progress and its response to treatment.

Clearly, the predominant factor in skin temperature is blood perfusion. Therefore, diseases that modify circulation, particularly in the extremities, may produce clear changes that can be shown by thermography. Generally, the skin or local temperature is increased by increased vascularity and decreased in ischemia or vascular restriction. A local hypothermic state may be the result of general body cooling or nervous tension. Smoking, arterial obstruction, or cysts may all result in this reduced temperature. A local temperature increase may be the result of inflammation, an arteriovenous fistula, or the effects of alcohol or exercise. Surgical procedures may include temperature changes, as in the case of chemical sympathectomy, in which peripheral vasodilation can be induced by selective blocking of the appropriate ganglion in the spine (see Figure 3.6). Abnormal temperatures may occur in the extremities in algodystrophy, and high localized temperatures over the site of a healing fracture have been used to monitor the repair process.

9.4.2. Vascular Disease

The thermographic evidence of vascular disease has been well documented by Winsor and Winsor (1976) and certain other authors. Arterial obstruction has been reported to produce areas of cold at particular sites. An internal carotid occlusion may show up as a cold area over the medial aspect of the eyebrow. The common carotid artery, if occluded, can produce distinct areas of cold over the lateral aspect of the forehead and face. There may be a loss of temperature contrast around the mouth (Figure 3.7). The radial and ulnar areas may be directly affected by occlusion of their specific blood supply. Occlusion of the iliac, common femoral, or popliteal artery has been reported to create abnormally cold areas, usually about 10 cm distal to the site of the blockage. Interruption of the digital artery can affect the whole of the digits, whereas occlusion of the arteriolar or capillary system may only lead to ischemia of the tips of the fingers or toes.

Hot localized areas on the legs are commonly seen, resulting from incompetent perforating veins. If the skin is well cooled and a light tourniquet is applied above and below the site, mild muscular contraction is sufficient to show the leakage of warm blood to the surrounding area. Other methods are used more widely than thermography for this investigation, but results have shown that thermography is superior to clinical judgment, and is less invasive than the alternative techniques so widely used.

Thermography is also capable of detecting deep vein thrombosis. It is particularly useful in pregnancy, where invasive methods are less acceptable. A diffuse area of increased temperature over the site of the collateral circulation may be identified. This may be more obvious in the calf after exercise. Peripheral circulation disorders associated with collagen diseases, systemic lupus erythematosus, scleroderma, and periarteritis nodosa may be indicated. Dynamic stress testing for vasospastic diseases such as Raynaud's syndrome is particularly appropriate with infrared radiometry or thermography. The possible side effects of certain drugs in exacerbating or inducing vasospastic disorders makes this a profitable area for further study.

9.4.3. Malignant Disease

The question of localized hyperthermia in relation to cancer is very complex. This one application probably accounts for some 60% of all clinical studies using thermography in the past twenty years. In some cases "hot" tumors have been located in the breast. A number of large-scale surveys in different countries have demonstrated that in breast disease the temperature patterns may be so complex that accurate separation of malignant from normal healthy tissue can be difficult. Many feel that thermography has a place in this area as part of a series of investigations. In breast cancer screening criticism of the so-called false positives recorded by individual criteria would appear to be losing ground. Further work will show if the observation can be substantiated that a physiological disturbance causing an "abnormal" thermogram may ultimately reveal malignancy. In general the relation between degree of hyperthermia and prognosis holds. However, this is of low diagnostic value. The possibility of monitoring conservative cancer therapy by this technique seems underused at present.

9.4.4. Osteoarticular Disease

Monitoring drug therapy in arthritis has become of increased importance in recent years. Chronic inflammation in the peripheral joints remains an assessment problem to the rheumatologist. The classical symptoms of inflammation are pain, swelling, redness, heat, and loss of function. Many of the subjective assessment methods currently used are based on pain, swelling, or loss of function. Antiinflammatory drugs of many different forms have been used in clinical trials in all the major countries. In some cases thermography has been employed as an objective measure of joint temperature. The reduction following therapy is most striking with locally injected drugs, usually of the steroid type, in the knee joint. However, oral drugs, aspirin derivatives, and the antirheumatic drugs, including cytotoxic agents, can also influence the thermographic image. Quantitation through the use of a thermographic index provides the clinician with direct information on the thermal state of the peripheral joints. As in all physiological measurements, however, due consideration to possible alternative reasons for temperature change and strict control of technique are important. Furthermore, no figure, when quoted, stands as an absolute, but only as a relative indicator of clinical state. Determination of changes in measurable parameters within the same patient, with careful use of placebos or control data, is essential.

Malignant diseases of the skeleton, metabolic disturbances, and Paget's disease of bone can, if the site is near the body surface, be effectively monitored by thermography. Early findings with microwave imaging offer promise for the study of spinal and skull lesions, which are difficult to observe with infrared thermography.

9.4.5. Tissue Viability

In all applications of medical thermography, the ability of the skin to lose heat by radiation is fundamental. Therefore, certain clinical problems in which skin injury or destruction occurs may also be appropriately monitored. A number of authors have drawn attention to the advantages of thermal imaging in cases of severe burns. For example, the extent of a burn injury is not always clearly defined for plastic surgery. Unfortunately many centers have dismissed this thermographic technique, because the scanning equipment is too bulky for use in the emergency room. However, modern instrumentation has greatly reduced size and increased mobility. In this instance, as in the measurement of corneal temperature, remote sensing is probably the only satisfactory technique available. Other applications in which thermography has been successfully used include ulcer and wound healing and the early detection of pressure sores and their management. Less expensive, portable imaging systems are adequate for these applications. Such systems are by now well-developed for field military use, but have been slow to reach clinical medicine. The full use of thermal imaging and body temperature measurement as a means of studying dynamic and physiological changes has yet to be realized.

APPENDICES

Common Causes of Increased Temperatures of Upper Limbs

Elbow
    Arthritis
    Trauma
    Infection
    Dermatitis
    Sarcoma

Wrist
    Arthritis
    Trauma
    Infection

Dorsal hand
    Reactive hyperemia
        Cold stress
        Warm stress
        Alcohol
    Tenosynovitis
    Dermatitis
    Bone fracture
    Infection
    Trauma

Metacarpal joints
    Arthritis, rheumatoid

Fingers
    Reactive hyperemia
    Algodystrophy
    Dermatitis
    Fracture
    Osteoarthritis (distal intraphalangeal joints)

Common Causes of Decreased Temperatures of Upper Limbs

Elbow
  Neurogenic lesions

Wrist
  Neurogenic lesions

Dorsal hand
  Arterial occlusion
  Diabetes
  Effects of smoking
  Old skin lesions/scar tissue

Fingers
  Vasospastic disorders
  Acrocyanosis
  Diabetic neuropathy
  Arteriosclerosis (early stages)
  Disseminated lupus
  Carpal tunnel syndrome
  Sudek's algodystrophy
  Dermatomyositis
  Scleroderma
  Vasospasms from smoking

Common Causes of Increased Temperatures of Lower Limbs

Upper leg
  Femoral thrombophlebitis
  Varices
  Trauma
  Vascularization (wound healing)

Knee joint
  Arthritis
    Rheumatoid
    Osteo- (inflammatory)
    Septic
  Acute synovitis (traumatic)
  Chondromalacia patellae
  Prepatellar bursitis
    Trauma
    Septic
  Gout

Lower leg
  Varices
  Paget's disease (tibia)
  Trauma

Fracture
Localized infection
Osteomyelitis
Sarcoma
Melanoma
Ruptured Baker's cyst
Deep vein thrombosis (collateral circulation)
Chronic venous insufficiency
Dermatological lesions (cellulitis)
Wound healing

Ankle joint
Arthritis
Sprain
Fracture
Infection
Postsympathectomy

Toes
Gout (especially first metatarsal phalangeal joint)
Infection
Dermatological lesions
Phlebitis

F

Common Causes of Decreased Temperatures of Lower Limbs

Upper leg
  Femoroiliac artery occlusion
  Full-thickness burns

Knee
  Ischemia
    Diabetic
    Popliteal artery occlusion
    Superficial femoral artery occlusion
  Localized neurological lesion
  Baker's cyst (knee/calf)
  Paralysis, hemi-/paraplegia
  Edema

Ankle
  Skin lesions
  Popliteal artery occlusion

Toes
  Digital artery ischemia
  Burgher's syndrome
  Acrocyanosis

Bibliography and References

Aarts, N. J. M., Gautherie, M., and Ring, E. F. J. (eds.). Thermography. *Bibl. Radiol.* 1975, *6*, 77–99.

Adolph, E. F. *Physiology of man in the desert.* Interscience, New York, 1947.

Afonso, S., Rowe, G. G., Castillo, C. A., and Crumpton, C. W. Intravascular and intracardiac blood temperature in man. *J. Appl. Physiol.* 1962, *17*, 706–708.

Aikas, E., and Piironen, P. Thermal exchanges of the human body in extreme heat. Technical Report AMRL-TDR 63-86. Wright-Patterson Air Force Base, 1963.

Anliker, M., and Friedli, P. Evaluation of high resolution thermograms by on-line digital mapping and color coding. *Appl. Radiol.* 1976, *5*, 114–118.

Atkins, E. Pathogenesis of fever. *Physiol. Rev.* 1960, *40*, 580–648.

Bargeton, D., Durand, J., Mensch-Dechene, J., and Decaud, J. Echanges de chaleur de la main. Rôle des réactions circulatoires et des variations de la température locale du sang artériel. *J. Physiol. (Paris)* 1956, *36*, 128–144.

Baumann, I. R., and Bligh, J. The influence of ambient temperature on drug-induced disturbances of body temperature. In *Temperature regulation and drug action* (pp. 241–251) (P. Lomax, E. Schönbaum, and J. Jacob, eds.). Karger, Basel, 1975.

Bedford, T. The effective radiative surface of the human body. *J. Hygiene* 1935, *35*, 303–306.

Bedford, T. *Basic principles of ventilation and heating.* Lewis, London, 1964.

Benzinger, T. H. Heat regulation: Homeostasis of central temperature. *Physiol. Rev.* 1969, *49*, 671–759.

Bianchi, S. D., Gatti, G., and Mecozzi, B. Circadian variations in the cutaneous thermal map in normal subjects. *Acta Thermogr.* 1974, *4*, 95–98.

Bligh, J. The thermosensitivity of the hypothalamus and thermoregulation in mammals. *Biol. Rev.* 1966, *41*, 317–367.

Bligh, J., and Johnson, K. G. Glossary of terms for thermal physiology. *J. Appl. Physiol.* 1973, *35*, 941–961.

Bligh, J., and Moore, R. E. (eds.). *Essays on temperature regulation.* North-Holland, Amsterdam, 1972.

Bohnenkamp, H., and Ernst, H. W. Untersuchungen zu den Grundlagen des Energie und Stoffwechsels. *Arch. Ges. Physiol.* 1931, *228*, 63–99.

Boutelier, C. Survie et protection des équipages en cas d'immersion accidentelle en eau froide. NATO Publication AGARD-AG 211, 1979.

Brand, P. W. Thermography in orthopaedics and in experimental stress. In *Medical thermography: Theory and clinical applications* (S. Uematsu, ed.). Brentwood, Los Angeles, 1976.

Brebner, D. F., and Kerslake, D. McK. The effect of cooling the legs on the rate of sweat production from the forearm when the rest of the body is not exchanging heat. *J. Physiol.* 1960, *152*, 65.

Brown, B. H., Bygrave, C., Robinson, P., and Henderson, H. P. A critique of the use of a thermal clearance probe for the measurement of skin blood flow. *Clin. Phys. Physiol. Meas.* 1980, *1*, 237–241.

Bruck, K. Thermoregulation: Control mechanisms and neural processes. In *Temperature regulation and energy metabolism in the new-born* (pp. 157–185) (J. C. Sinclair, ed.). Grune and Stratton, New York, 1978.

Budd, G. M. Physiological research at Australian stations in the Antarctic and sub-Antarctic. In *Antarctic research series*, Volume 22 (pp. 27–54) (E. K. E. Gunderson, ed.). American Geophysical Union, Washington, D.C., 1974.

Buettner, K. Warme Übertragung durch Leitung und Convektion Verdünstung und Strahlung in Bioklimatologie und Meteorologie. *Veroeff. Truess. Meteorol. Inst.* 1935, *10.5*, 1–137.

Buettner, K. Diffusion of water and water vapor through human skin. *J. Appl. Physiol.* 1953, *6*, 229–233.

Burton, A. C., and Edholm, O. G. *Man in a cold environment.* Arnold, London, 1955.

Cabanac, M., and Hardy, J. D. Réponses unitaires et thermorégulatrices lors de réchauffements et refroidissements localisés de la région pré-optique et du mésencéphale. *J. Physiol. (Paris)* 1969, *61*, 331–347.

Cabanac, M., Stolwijk, J. A. J., and Hardy, J. D. Effect of temperature and pyrogens on single unit activity in rabbit's brain stem. *J. Appl. Physiol.* 1968, *24*, 645–652.

Cage, G. W., and Dobson, R. L. Sodium secretion and reabsorption in the human eccrine sweat gland. *J. Clin. Invest.* 1965, *44*, 1270–1276.

Cena, K., and Clark, J. A. (eds.). *Bioengineering, thermal physiology, and comfort.* Elsevier, Amsterdam, 1981.

Chappuis, P., Pittet, P., and Jequier, E. Heat storage regulation in exercise during thermal transient. *J. Appl. Physiol.* 1976, *40*, 384–392.

Chatonnet, J. Sur l'origine de la source de chaleur libérée dans la régulation chimique de la température. *J. Physiol. (Paris)* 1959, *51*, 319–378.

Clark, J. A. Effects of surface emissivity and viewing angle on errors in thermography. *Acta Thermogr.* 1976, *1*, 138–141.

Clark, J. A., and Cena, K. Solar and thermal radiative heat loads in the energy balance of man. *Eng. Med.* 1976, *5*, 75–78.

Clark, R. P. Human skin temperature and convective heat loss. In *Bioengineering, thermal physiology, and comfort* (pp. 57–76) (K. Cena and J. A. Clark, eds.). Elsevier, Amsterdam, 1981.

Colin, G., Boutelier, C., and Houdas, Y. Détermination expérimentale de la surface effective de radiation thermique chez l'homme. *C. R. Acad. Sci.* 1966, *282*, 1966–1969.

Colin, J., and Houdas, Y. Experimental determination of the coefficient of heat exchange of the human body. *J. Appl. Physiol.* 1967a, *22*, 31–38.

Colin, J., and Houdas, Y. Mécanismes de la perspiration insensible cutanée. *Pathol. Biol.* 1967b, *15*, 5–13.

Colin, J., Timbal, J., Boutelier, C., Guieu, J. D., and Houdas, Y. Combined effect of radiation and convection. In *Physiological and behavioral temperature regulation* (pp. 81–96) (J. D. Hardy, A. P. Gagge, and J. A. J. Stolwijk, eds.). C. C. Thomas, Springfield, 1970.

Colin, J., Timbal, J., Houdas, J., Boutelier, C., and Guieu, J. D. Computation of mean body temperature from rectal and skin temperatures. *J. Appl. Physiol.* 1971, *31*, 484–489.

Collins, K. J., and Weiner, J. S. Endocrinological aspects of exposure to high environmental temperatures. *Physiol. Rev.* 1968, *48*, 785–839.

Cooper, K. E., and Kerslake, D. McK. Changes in heart rate during exposure of the skin to radiant heat. *Clin. Sci.* 1955, *14*, 125–135.

Craigh, A. B., and Dvorak, M. Thermal regulation during water immersion. *J. Appl. Physiol.* 1966, *21*, 1577–1585.

Dereniak, E. L. Thermographic instrumentation. In *Imaging for medicine*, Volume 1 (pp. 419–441) (S. Nudelman and D. D. Patton, eds.). Plenum Press, New York, 1981.

Edelman, I. Thyroid thermogenesis. *N. Engl. J. Med.* 1974, *56*, 1303–1308.

Edholm, O. G., and Bacharach, A. L. *The physiology of human survival.* Academic Press, New York, 1965.

Edholm, O. G., and Weiner, J. S. *Principles and practice of human physiology.* Academic Press, London, 1981.

Eisenmann, J. S. Unit activity studies of thermoreceptive neurons. In *Essays on temperature regulation* (pp. 55–69) (J. Bligh and R. E. Moore, eds.). North-Holland, Amsterdam, 1972.

Fanger, P. O. *Thermal comfort.* Danish Technical Press, Copenhagen, 1970.

Feldberg, W. On the mechanism of action of pyrogens. In *Pyrogens and fever* (pp. 115–129) (G. E. W. Wolstenholme and J. Birch, eds.). Churchill Livingstone, Edinburgh, 1971.

Fergason, J. L. Liquid crystals. *Sci. Am.* 1964, *211*, 77–85.

Folk, G. E., Jr. Climatic change and acclimatisation. In *Bioengineering, thermal physiology, and comfort* (K. Cena and J. A. Clark, eds.). Elsevier, Amsterdam, 1981.

Fox, R. H. Temperature regulation with special reference to man. In *Recent advances in physiology* (pp. 340–405) (J. Linden, ed.). Churchill Livingstone, Edinburgh and London, 1974.

Fox, R. H., and Hilton, S. M. Bradykinin formation in human skin as a factor in heat vasodilatation. *J. Physiol. (London)* 1958, *142*, 219–232.

Fox, R. H., Goldsmith, R., Kidd, D. J., and Lewis, H. E. Acclimatisation to heat in man by controlled elevation of body temperature. *J. Physiol. (London)* 1963, *166*, 530–547.

Fox, R. H., Davies, T. W., Marsh, F. P., and Urich, H. Hypothermia in a young man with an anterior hypothalamic lesion. *Lancet* 1970, *2*, 185–188.

Gagge, A. P. A new physiological variable associated with sensible and insensible perspiration. *Am. J. Physiol.* 1937, *120*, 277–287.

Gagge, A. P. The effective radiant flux. An independent variable that describes the physical effect of thermal radiation on man. In *Physiological and behavioral temperature regulation* (pp. 34–45) (J. D. Hardy, A. P. Gagge, and J. A. J. Stolwijk, eds.). C. C. Thomas, Springfield, 1970.

Gagge, A. P., Hardy, J. D., and Stolwijk, J. A. J. Proposed standard system of symbols for thermal physiology. *J. Appl. Physiol.* 1969, *10*, 547–569.

Gale, C. C. Neuroendocrine aspects of thermoregulation. *Annu. Rev. Physiol.* 1973, *35*, 391–430.

Gautherie, M. Etude par thermographie infrarouge des propriétés thermiques de tissu humain "in vivo." *Rev. Fr. Etud. Clin. Biol.* 1969, *14*, 885–901.

Gee, G. K., and Goldman, R. F. Heat loss of man in total water immersion. *Physiologist* 1973, *16*, 318.

Goldman, R. F., Green, E., and Iampetro, P. F. Tolerance of hot wet environments by resting men. *J. Appl. Physiol.* 1965, *20*, 271–277.

Goldsmith, R. Acclimatisation to heat and cold in man. In *Recent advances in medicine,* Volume 17 (pp. 299–321) (D. N. Baron, N. Compston, and A. M. Dawson, eds.). Churchill Livingstone, Edinburgh, 1977.

Gonzalez, R. R. Exercise physiology and sensory responses. In *Bioengineering, thermal physiology, and comfort* (pp. 123–144) (K. Cena and J. A. Clark, eds.). Elsevier, Amsterdam, 1981.

Gordon, R. A. (ed.). *International symposium on malignant hyperthermia.* C. C. Thomas, Springfield, 1973.

Graf, W. Patterns of human liver temperature. *Acta Physiol. Scand.* 1959, *46* (Suppl. 160), 1–135.

Greenleaf, J. E. Blood electrolytes and exercise in relation to temperature regulation in man. In *The pharmacology of thermoregulation* (pp. 72–84) (E. Schonbaum and P. Lomax, eds.). Karger, Basel, 1973.

Greenleaf, J. E., and Castle, B. L. Exercise temperature regulation in man during hypohydration and hyperhydration. *J. Appl. Physiol.* 1971, *30*, 847–853.

Guerin, H. *Traité de manipulation et d'analyse des gaz.* Masson, Paris, 1952.

Guibert, A., and Taylor, C. L. Radiation area of the human body. *J. Appl. Physiol.* 1952, *5*, 24–37.

Guieu, J.-D., and Hardy, J. D. Effects of heating and cooling the spinal cord on preoptic unit activity. *J. Appl. Physiol.* 1970a, *29*, 675–683.

Guieu, J.-D., and Hardy, J. D. Integrative activity of preoptic units. I. Response to local and peripheral temperature changes. *J. Physiol. (Paris)* 1970b, *63*, 253–256.

Hackett, M. E. J. The place of thermography in medicine. *Acta Thermogr.* 1976, *1*, 176–180.

Halberg, F. Laboratory techniques and rhythmometry. In *Biological aspects of circadian rhythms* (pp. 1–26) (J. N. Mills, ed.). Plenum Press, London, 1973.

Hammel, H. T. Regulation of internal body temperature. *Annu. Rev. Physiol.* 1968, *30*, 641–710.

Hancock, P. A. Heat stress impairment of mental performance: A revision of tolerance limits. *Aviat. Space Environ. Med.* 1981, *52*, 177–180.

Hardy, J. D. The radiation of heat from the human body. *J. Clin. Invest.* 1934, *13*, 539–615.

Hardy, J. D. Physiology of temperature regulation. *Physiol. Rev.* 1961, *41*, 521–606.

Hardy, J. D. (ed.). *Temperature: Its measurement and control in science and industry*, Volume III. Part 3: *Biology and medicine.* Reinhold, New York, 1963.

Hardy, J. D. Central and peripheral factors in physiological temperature regulation. In *Les concepts de Claude Bernard sur le milieu intérieur* (pp. 247–283). Masson, Paris, 1967.

Hardy, J. D., and Dubois, E. F. Basal metabolism, radiation, convection, and evaporation at temperatures from 22 to 35°C. *J. Nutr.* 1938, *15*, 477–492.

Hardy, J. D., Dubois, E. F., and Soderström, G. F. The technique of measuring radiation and convection. *J. Nutr.* 1938, *15*, 461.

Hardy, J. D., Hellon, R. F., and Sutherland, K. Temperature-sensitive neurons in the dog's hypothalamus. *J. Physiol. (London)* 1964, *175*, 242–253.

Hardy, J. D., Gagge, A. P., and Stolwijk, J. A. J. (eds). *Physiological and behavioral temperature regulation.* C. C. Thomas, Springfield, 1970.

Hellon, R. Hypothalamic neurons responding to changes in hypothalamic and ambient temperatures. In *Physiological and behavioral temperature regulation* (pp. 463–471) (J. D. Hardy, A. P. Gagge, and J. A. J. Stolwijk, eds.). C. C. Thomas, Springfield, 1970.

Hemingway, A. Shivering. *Physiol. Rev.* 1963, *43*, 397–422.

Henane, R. La dépression sudorale au cours de l'hyperthermie contrôlée chez l'homme. *J. Physiol. (Paris)* 1972, *64*, 147–163.

Hensel, H. Neural processes in thermoregulation. *Physiol. Rev.* 1973, *53*, 948–1017.

Hensel, H. *Thermoregulation and temperature regulation.* Academic Press, London, 1981.

Hertig, B. A. Human physiological responses to heat stress: Males and females compared. *J. Physiol. (Paris)* 1970, *63*, 269–273.

Hertzmann, A. B. Individual differences in regional sweating. *J. Appl. Physiol.* 1957, *10*, 242–248.

Hertzmann, A. B. Vasomotor regulation of cutaneous circulation. *Physiol. Rev.* 1959, *39*, 280–306.

Holti, G., and Mitchell, K. W. Estimation of the nutrient skin blood flow using a segmental thermal clearance probe. *Clin. Exp. Dermatol.* 1978, *3*, 189–198.

Horvath, S. M. Man in cold stress. In *Advances in climatic physiology* (pp. 172–177) (S. Itoh, K. Ogata, and H. Yoshimura, eds.). Igaku Shoin, Tokyo, 1972.

Horwitz, B. A., and Smith, R. E. Function and control of brown fat thermogenesis during cold exposure. In *Bioenergetics* (pp. 134–140) (R. E. Smith, J. P. Hannon, J. L. Shields, and B. A. Horwitz, eds.). FASEB, Washington, D.C., 1972.

Houdas, Y., and Guieu, J. D. *Physiologie humaine*, Volume VI: *La fonction thermique*. SIMEP, Lyon-Villerubanne, 1977.

Houdas, Y., and Guieu, J. D. (eds.) *New trends in thermal physiology*. Masson, Paris, 1978.

Houdas, Y., and Sauvage, A. La réponse du thermostat humain á une entreé pente de la charge thermique externe. *J. Physiol. (Paris)* 1971, *63*, 279–281.

Houdas, Y., Sauvage, A., Bonaventure, M., and Guieu, J. D. Control of heat exchange: An alternate concept for temperature regulation. In *Regulation and control in physiological systems* (pp. 217–220) (A. S. Iberall and A. C. Guyton, eds.). International Federation for Automatic Control, Pittsburgh, 1973a.

Houdas, Y., Sauvage, A., Bonaventure, M., Ledru, C., and Guieu, J.-D. Thermal control in man: Regulation of central temperature or adjustments of heat exchanges by servo-mechanism? *J. Dyn. Syst. Meas. Control* 1973b, *95* (series G), 331–335.

Hull, D., and Smales, O. R. C. Heat production in the newborn. In *Temperature regulation and energy metabolism in the newborn* (pp. 129–156) (J. C. Sinclair, ed.). Grune and Stratton, New York, 1978.

Iggo, A. Cutaneous thermoreceptors in primates and sub-primates. *J. Physiol. (London)* 1969, *200*, 403–430.

Ingram, D. L., and Mount, L. E. Man and animals in hot environments. Springer Verlag, New York, 1975.

Itoh, S., Ogata, K., and Yoshimura, H. (eds.). *Advances in climatic physiology*. Igaku Shoin, Tokyo, 1972.

Jansky, L. *Depressed metabolism and cold thermogenesis*. Charles University Press, Prague, 1975.

Jequier, E., Dolivo, M., and Vanotti, A. Temperature regulation in exercise: The characteristics of proportional control. *J. Physiol. (Paris)* 1970, *63*, 303–305.

Jones, C. H. Physical aspects of thermography in relation to clinical techniques. *Bibl. Radiol.* 1975, *6*, 1–8.

Kaciuba-Uscilko, H., Brzezinska, Z., and Kobryn, A. Metabolic and temperature responses to physical exercise in thyroidectomized dogs. *Eur. J. Appl. Physiol.* 1979, *40*, 219–226.

Keatinge, W. R. *Survival in cold water*. Blackwell, Oxford, 1969.

Kerslake, D. McK. *The stress of hot environments*. Cambridge University Press, Cambridge, 1972.

Kerslake, D. McK., and Cooper, K. E. Vasodilatation in the hand in response to heating the skin elsewhere. *Clin. Sci.* 1950, *9*, 32–47.

Kluger, M. J. Temperature regulation, fever and disease. *Int. Rev. Physiol. (Environ. Physiol.* III), 1979, *20*, 209–251.

Klussmann, F. W., and Pierau, F. K. Extrahypothalamic deep body thermosensitivity. In *Essays on temperature regulation* (pp. 87–104) (J. Bligh and R. E. Moore, eds.). North-Holland, Amsterdam, 1972.

Kozlowski, S., Kaciuba-Uscilko, H., Greenleaf, J. E., and Brzezinska, Z. The effect of thyroxin on temperature regulation during physical exercise in dogs. In *Temperature regulation and drug action* (pp. 361–366) (P. Lomax, E. Schonbaum, and J. Jacob, eds.). Karger, Basel, 1975.

Kreith, F. *Transmission de chaleur et thermodynamique.* Masson, Paris, 1967.

Kuno, Y. *The physiology of human perspiration.* Churchill, London, 1956.

Leblanc, J. *Man in the cold* (American Lecture Series No. 986). C. C. Thomas, Springfield, 1975.

Leblanc, J. L'adaptation de l'homme au froid. *Rev. Med.* 1976, *32*, 1683–1690.

Lecroart, J. L. Contribution à l'étude des échanges thermiques chez l'homme: Détermination d'une méthode de référence pour la mesure des températures cutanées moyennes par thermométrie infra-rouge. Thesis, University of Lille, 1980.

Leithead, C. S., and Lind, A. R. *Heat stress and heat disorders.* Cassell, London, 1964.

Liebermeister, C. *Handbuch der Pathologie und Therapie des Fiebers.* Vogel, Leipzig, 1875.

Lind, A. R. A physiological criterion for setting thermal environmental limits for everyday work. *J. Appl. Physiol.* 1963, *18*, 51–56.

Lipkin, M., and Hardy, J. D. Measurement of some thermal properties of human tissues. *J. Appl. Physiol.* 1954, *7*, 212–217.

Lloyd, D. P. C. Secretion and reabsorption in sweat glands. *Proc. Natl. Acad. Sci.* 1959, *45*, 405–409.

Lomax, P., Schonbaum, E., and Jacob, J. (eds.). *Temperature regulation and drug action.* Karger, Basel, 1975.

Mitchell, D., and Wyndham, C. H. Comparison of weighting formulas for calculating mean skin temperature. *J. Appl. Physiol.* 1969, *26*, 616–622.

Mitchell, D., Wyndham, C. H., Vermeulen, A. J., Hogdson, T., Atkins, A. R., and Hofmeyr, H. S. Radiant and convective heat transfer of men in dry air. *J. Appl. Physiol.* 1969, *26*, 111–118.

Mitchell, D., Atkins, A. R., and Wyndham, C. H. Mathematical and physical models of thermoregulation. In *Essays on temperature regulation* (pp. 37–54) (J. Bligh and R. E. Moore, eds.). North-Holland, Amsterdam, 1972.

Molnar, G. W. Survival of hypothermia by men immersed in the ocean. *JAMA* 1946, *131*, 1046–1050.

Morehouse, L. E., and Miller, A. T. *Physiology of exercise.* C. V. Mosby, St. Louis, 1963.

Myers, R. D. Hypothalamic mechanisms of pyrogen action in the cat and monkey.

In *Pyrogens and fever* (pp. 131–153) (G. E. W. Wolstenholme and J. Birch, eds.). Churchill Livingstone, Edinburgh, 1971.

Myers, R. D. An integrative modal of monoamine and ionic mechanisms in the hypothalamic control of body temperature. In *Temperature regulation and drug action* (pp. 32–42) (P. Lomax, E. Schonbaum, and J. Jacob, eds.). Karger, Basel, 1975.

Nadel, E. R. (ed.). *Problems with temperature regulation during exercise.* Academic Press, New York, 1977.

Nadel, E. R. Temperature regulation during exercise. In *New trends in thermal physiology* (pp. 143–153) (Y. Houdas and J.-D. Guieu, eds.). Masson, Paris, 1978.

Nadel, E. R., Pandolf, K. B., Roberts, M. F., and Stolwijk, J. A. J. Mechanisms of thermal acclimation to exercise and heat. *J. Appl. Physiol.* 1974, *30*, 847–753.

Nakayama, T. Thermosensitive neurons in the brain. In *Advances in climatic physiology* (pp. 68–76) (S. Itok, K. Ogata, and H. Yoshimura, eds.). Igaku Shoin, Tokyo, 1972.

Nelson, N. A., Eichna, L. W., Horvath, S. M., Shelley, W. B., and Hatch, T. F. Thermal exchanges of man at high temperatures. *Am. J. Physiol.* 1947, *151*, 626–652.

Nielsen, B. Thermoregulation in rest and exercise. *Acta Physiol. Scand.* 1969, *76*, (suppl. 323), 1–74.

Nishi, Y., and Gagge, A. P. Direct evaluation of convective heat transfer coefficient by naphthalene sublimation. *J. Appl. Physiol.* 1970, *29*, 830–838.

Noel, P., Hubert, J. P., Ectors, M., Franken, L., and Flament Durand, J. Agenesis of the corpus callosum associated with relapsing hypothermia. *Brain* 1973, *96*, 359–368.

Ochi, J., and Sano, Y. Morphological attempts to solve some unsettled problems on the human eccrine sweat glands. In *Advances in climatic physiology* (pp. 79–92) (S. Itoh, K. Ogata, and H. Ishimura, eds.). Igaku Shoin, Tokyo, 1972.

Ogawa, T. Local determinants of sweat gland activity. In *Advances in climatic physiology* (pp. 93–108) (S. Itoh, K. Ogata, and H. Yoshimura, eds.). Igaku Shoin, Tokyo, 1972.

Onoe, M. Preston, K., and Rosenfeld, A. (eds.). *Real-Time Medical Image Processing.* Plenum Press, New York, 1980.

Palmes, E. D., and Park, C. R. Thermal regulation during early acclimatization to work in a hot dry environment. Technical Report 2-17-1, Medical Department, Field Research Laboratory, Fort Knox, 1947.

Paolone, A. M., Wells, C. L., and Kelly, G. T. Sexual variations in thermoregulation during heat stress. *Aviat. Space Environ. Med.* 1978, *49*, 715–719.

Ramanathan, N. L. A new weighting system for mean surface temperature of the human body. *J. Appl. Physiol.* 1964, *19*, 531–533.

Rapp, G. M. Convective mass transfer as the coefficient of evaporative heat loss from the human skin. In *Physiological and behavioral temperature regulation* (pp.

55–80) (J. D. Hardy, A. P. Gagge, and J. A. J. Stolwijk, eds.). C. C. Thomas, Springfield, 1970.

Rapp, G. M. Convection coefficients of man in a forensic area of thermal physiology: Heat transfer in underwater exercise. *J. Physiol. (Paris)* 1971, *63*, 392–396.

Randall, W. C. The physiology of sweating. *Am. J. Physiol. Med.* 1953, *32*, 292–318.

Rawson, R. O., and Quick, K. P. Localization of intra-abdominal thermoreceptors in the ewe. *J. Physiol. (London)* 1972, *222*, 665–677.

Ring, E. F. J. Computer processing of infrared thermograms applied to bone and joint pathology. *Optonics and Photonics Applied to Medicine* 1979, *211*, 141–144.

Robinson, S. Circulatory adjustments of men in hot environments. In *Temperature: Its measurement and control in science and industry*, Volume III. Part 3: *Biology and medicine* (pp. 287–297) (J. D. Hardy, ed.). Reinhold, New York, 1963.

Robinson, S., and Robinson, A. H. Chemical composition of sweat. *Physiol. Rev.* 1954, *34*, 202–220.

Rowell, L. B. Human cardiovascular adjustments to exercise and thermal stress. *Physiol. Rev.* 1974, *54*, 75–159.

Saltin, B., and Hermansen, L. Esophageal, rectal, and muscle temperature during exercise. *J. Appl. Physiol.* 1966, *21*, 1757–1762.

Saltin, B., Gagge, A., and Stolwijk, J. Muscle temperature during sub-maximal exercise in man. *J. Appl. Physiol.* 1968, *25*, 679–688.

Sato, K. Current knowledge on the energy metabolism and the secretory mechanism of the eccrine sweat glands. In *Secretory mechanism of exocrine glands* (pp. 588–607) (N. A. Thorn and O. H. Petersen, eds.). Munksgaard, Copenhagen, 1974.

Sato, K. The physiology, pharmacology and biochemistry of the eccrine sweat gland. *Rev. Physiol. Biochem. Pharmacol.* 1977, *79*, 51–131.

Shibolet, S., Lancaster, M. C., and Danon, Y. Heat stroke: A review. *Aviat. Space Environ. Physiol.* 1976, *47*, 280–301.

Shvartz, E., Bhattacharya, A. Sperinde, S. J., Brock, P. J., Sciaraffa, D., and Van Beaumont, W. Sweating responses during heat acclimation and moderate conditioning. *J. Appl. Physiol. (Resp. Environ. Exercise Physiol.)* 1979, *46*, 675–680.

Simon, E. Temperature regulation: The spinal cord as a site of extra-hypothalamic thermoregulatory functions. *Rev. Physiol. Biochem. Pharmacol.* 1974, *71*, 1–75.

Simon, E., and Jriki, M. Sensory transmission of spinal heat and cold sensitivity in ascending spinal neurons. *Pflügers Arch. Ges. Physiol.* 1971, *328*, 103–120.

Sinclair, J. C. (ed.). *Temperature regulation and energy metabolism in the newborn.* Grune and Stratton, New York, 1978.

Smith, R. E., Hannon, J. P., Shields, J. L., and Horwitz, B. A. (eds). *Bioenergetics.* FASEB, Washington, D.C. 1972.

Staak, W. J. B. M. Van der. Experiences with a heated thermocouple. *Dermatologica* 1966, *132*, 192–205.

States, S. J. Weather and death in Pittsburgh, Pennsylvania: A comparison with

Birmingham, Alabama. *Int. J. Biometeorol.* 1977, *21*, 7–15.

Steketee, J. S. Spectral emissivity of skin and pericardium. *Phys. Med. Biol.* 1973, *18*, 686–694.

Steketee, J. The influence of the environment on infrared thermography. *Acta Thermogr.* 1979, *4*, 43–47.

Stolwijk, J. A. J. A mathematical model of physiological temperature regulation. NASA Contract Report No. 1855, pp. 184–265, 1971.

Stuart, D. G., Eldred, E., Hemingway, A., and Kawamura, Y. Neural regulation of the rhythm of shivering. In *Temperature: Its measurement and control in science and industry*, Volume III. Part 3: *Biology and medicine* (pp. 545–557) (J. D. Hardy, ed.). Rheinhold, New York, 1963.

Suzuki, M. Thyroid activity and cold adaptability. In *Advances in climatic physiology* (pp. 93–108) (S. Itoh, K. Ogata, and H. Yoshimura, eds.). Igaku Shoin, Tokyo, 1972.

Tanche, M., and Therminarias, A. Thyroxine and catecholamines during cold exposure in dogs. *Fed. Proc.* 1969, *28*, 1257–1261.

Tanche, M., Chatonnet, J., Guieu, J. D., and Couderc, P. Sur les voies médullaires nécessaires à la commande du frisson thermique. *C. R. Soc. Biol. (Paris)* 1960, *154*, 993–994.

Thauer, R. Mécanismes périphériques et centraux de la régulation de la température. *Arch. Sci. Physiol.* 1961, *15*, 95–123.

Thauer, R. Thermosensitivity of the spinal cord. In *Physiological and behavioral temperature regulation* (pp. 472–492) (J. D. Hardy, A. P. Gagge, and J. A. J. Stolwijk, eds.). C. C. Thomas, Springfield, 1970.

Thauer, R., and Simon, E. Spinal cord and temperature regulation. In *Advances in climatic physiology* (pp. 22–49) (S. Itoh, K. Ogata, and H. Yoshimura, eds.). Igaku Shoin, Tokyo, 1972.

Therminarias, A., Chirpaz, M. F., Lucas, A., and Tanche, M. Catecholamines in dogs during cold adaptation by repeated immersion. *J. Appl. Physiol. (Resp. Environm. Exercise Physiol.)* 1979, *46*, 662–668.

Thermographic terminology. *Acta Thermogr.* 1978, Suppl. 2 (E. F. J. Ring, Y. Houdas, and G. F. Pistolesi, eds.).

Thermography in Locomotor Diseases. Recommended Procedure. Locomotor Diseases Group Report of the Anglo-Dutch Thermographic Society. *Eur. J. Rheumatol. Inflamm.* 1979, *2*, 299–306.

Timbal, J., Colin, J., Guieu, J.-D., and Boutelier, C. A mathematical study of thermal losses by sweating in man. *J. Appl. Physiol.* 1969, *27*, 726–730.

Underwood, C. R., and Ward, E. J. The solar radiation area of man. *Ergonomics* 1966, *9*, 155–168.

Veale, W. L., and Cooper, K. E. Comparison of sites of action of prostaglandin E and leucocyte pyrogen in brain. In *Temperature regulation and drug action* (pp. 218–226) (P. Lomax, E. Schonbaum, and J. Jacob, eds.). Karger, Basel, 1975.

Vogt, J. J. Déclenchement et rendement évaporatoire de la sudation thermique. *Trav. Hum.* 1968, *31*, 165–174.

Watmough, D. J., and Oliver, R. Wavelength dependence of skin emissivity. *Physiol. Med. Biol.* 1969, *14*, 201–204.

Webb, P. Thermal stress in undersea activity. In *Underwater physiology V* (pp. 705–724) (C. J. Lambertsen, ed.). FASEB, Bethesda, 1976.

Webb, P., Annis, J. F., and Troutman, S. J. Heat flow regulation. In *New trends in thermal physiology* (pp. 29–32) (Y. Houdas and J.-D. Guieu, eds.). Masson, Paris, 1978.

Webster, A. J. Adaptation to cold. In *Environmental physiology* (pp. 71–106) (D. Robertshaw, ed.). Butterworths, London, 1974.

Wells, C. L., and Horvath, S. M. Response to exercise in a hot environment as related to the menstrual cycle. *J. Appl. Physiol.* 1974, *36*, 299–302.

Wenger, C. B., Roberts, M. F., Nadel, E. R., and Stolwijk, J. A. J. Thermoregulatory control of finger blood flow. *J. Appl. Physiol.* 1975, *38*, 1078–1082.

Werner, J. The concept of regulation for human body temperature. *J. Therm. Biol.* 1980, *5*, 75–82.

Wilkie, D. R. Thermodynamics and the interpretation of biological heat measurements. In *Progress in biophysics and biophysical chemistry* (pp. 260–298) (J. A. V. Butler and B. Katz, eds.). Academic Press, New York, 1960.

Winslow, C. E. A. Man's heat exchanges with his thermal environment. In *Temperature: Its measurement and control in science and industry*, Volume I (p. 509) (J. D. Hardy, ed.). Reinhold, New York, 1941.

Winslow, C. E. A., and Gagge, A. P. Influence of physical work on physiological reactions to the thermal environment. *Am. J. Physiol.* 1941, *134*, 664.

Winslow, C. E. A., and Herrington, L. P. *Temperature and human life.* Princeton University Press, Princeton, 1979.

Winsor, T., and Winsor, D. Thermography in cardiovascular disease. In *Medical thermography: Theory and clinical applications* (pp. 121–142) (S. Uematsu, ed.). Brentwood, Los Angeles, 1976.

Wit, A., and Wang, S. C. Temperature-sensitive neurons in preoptic–anterior hypothalamus region: Actions of pyrogen and acetylsalicylate. *Am. J. Physiol.* 1968, *215*, 1160–1169.

Wolstenholme, G. E. W., and Birch, J. (eds.). *Pyrogens and fever.* Churchill Livingstone, Edinburgh, 1971.

Wyndham, C. H. The physiology of exercise under heat stress. *Annu. Rev. Physiol.* 1973, *35*, 193–220.

Wyndham, C. H., Strydom, N. B., Morrison, J. F., Williams, C. G., Bredell, G. A. G., Maritz, J. S., and Murro, A. Criteria for physiological limits for work in heat. *J. Appl. Physiol.* 1965, *20*, 37–45.

Index